New Mining Tools and Methods for Roadheader Mining Heads

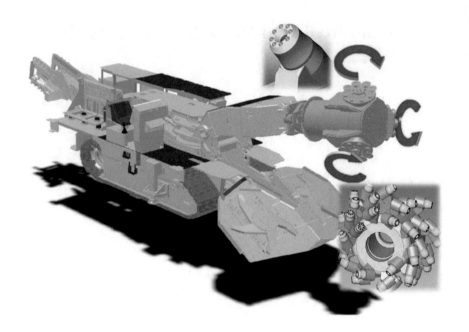

Krzysztof Kotwica

New Mining Tools
and Methods for Roadheader
Mining Heads

Springer

Krzysztof Kotwica
Department of Machinery Engineering
and Transport
AGH University of Science and Technology
Cracow, Poland

ISBN 978-3-030-96396-5   ISBN 978-3-030-96394-1   (eBook)
https://doi.org/10.1007/978-3-030-96394-1

The proposals of new mining tools, mining heads and mining techniques for roadheaders.

This Springer imprint is published by the registered company Springer Nature Switzerland AG
The registered company address is: Gewerbestrasse 11, 6330 Cham, Switzerland

Preface

Currently, in underground mines, one of the most used methods of hard rocks mining, in the process of row minerals excavation and drilling of exploratory and opening-out headings or tunnels, is the mechanical method. It is very often performed in rocks with very unfavorable parameters. The most used machines in underground mines are roadheaders, continuous miners and shearers. Cutting tools are used in these machines, mainly tangential-rotary (conical) and radial picks. In this case, the process of hard rock mining is very often associated with the generation of threats such as sparks, dust and increased wear of cutting tools.

Another mining technique that used the disk tools eliminated or decreased these threats, but it can be used only in specific geological mining conditions and mainly for a considerable runway of gallery or tunnel.

Actions taken to increase the durability of cutting tools, such as new geometrical and material solutions for the sintered carbide insert or the use of rings made of wear-resistant materials securing the place of fixing the carbide insert, do not give satisfactory results, especially in the case of complex and compact rock mining.

This problem was the reason for researching at AGH University of Science and Technology in Kraków on developing innovative solutions for cutting tools and holders, allowing for effective mining of compact rocks. As part of these works, a new solution of the mining head for roadheaders was also developed, in which asymmetrical mini-disk tools were used. The author took part in these works as the leading investigator or a research co-contractor.

The book consists of seven main chapters. Chapter 1 describes the methods, machines and devices currently used for mineral resources excavation and gallery driving. The benefits of mechanical technology were also presented.

Chapter 2 presents the problems occurring during the mechanical mining of hard rocks using cutting tools. Also, the methods currently used to limit these problems were described. Chapter 3 presents the possibilities and disadvantages of hard rock mining using disk tools with the mining method by static crumpling or back incision.

In Chap. 4, the new solution of a crown pick as an alternative for the conical pick and the result of research works with the use of new tools on the mining heads of roadheaders were presented.

In Chap. 5, the construction and work principle of a new solution of the lubricated holder for conical picks allowing for tool rotation was described. The results of research and laboratory, field and industrial tests were presented.

Chapter 6 presents the result of laboratory and field tests carried out in AGH-UST, Cracow, using the unconventional design of mining heads for roadheaders with mini asymmetrical disk tools of complex motion trajectory.

In the last Chap. 7, the further directions of future research were presented, allowing for improving the presented solutions and their wider use.

All the above-described solutions of tools, holders and mining heads can be successfully used for mining hard rock with roadheaders. It is crucial when the transformation of underground mining is realized. This book can be used by mining industry employees, manufacturers of mining tools, machines and equipment and employees and students of mining and mechanical faculties of technical universities.

To write this book, the author used some of the information and materials contained in publications of which he was the sole author or the principal co-author.

Cracow, Poland   Krzysztof Kotwica

Acknowledgements

This book arose from the results of the research works performed in the Department of Mining, Dressing and Transport Machines (currently Department of Machinery Engineering and Transport), AGH University of Science and Technology, Cracow, Poland, 1996–2016. These research works were carried out as a part of research projects financed in whole or in part by the Polish Committee for Scientific Research, the Polish National Science Centre and Polish National Centre for Research and Development, as well as research works financed by AGH University of Science and Technology, Cracow, Poland, as part of my own and statutory research.

I want to thank all the employees of the Department of Mining, Dressing and Transport Machines and workers from factories producing mining machinery and mines, participating in this research works, without whose efforts this could not have written book.

I want to thank the staff at Springer, particularly Anthony Doyle and Jayanthi Krishnamoorthi, for their help and support.

Cracow, Poland   Krzysztof Kotwica

Contents

Chapter 1
Introduction

Abstract This chapter briefly describes the methods, machines, and mining tools currently used for mineral resources excavation and gallery driving. In the first part, energetic aspects of the rock mining process are presented. Then, the methods used in the mining of hard rocks are discussed. In the next part, the mechanical method and method using explosives are compared, which allowed for the presentation of the benefits of mechanical technology. The last part describes the machines and tools used for rock mining.

Gallery driving is one of the leading underground mining operations performed to uncover mineral resources deposits and prepare them for exploitation. The gallery and tunnel driving processes are complex due to considerable difficulties and limitations of geological and mining and technical nature. These works are the most labour-intensive and time-consuming. The essential factors underlying the choice of the gallery driving techniques are [1–5]:

- the geological-mining conditions in which the heading is driven;
- the gallery's or tunnel's functionality in terms of how it will be used in the future;
- the gallery's or tunnel's length and cross-section, as well as the methods for roof supporting against rockfalls and convergence due to massive rock pressure;
- the time in which the work should be completed.

These factors mentioned above greatly influence the costs incurred to uncover deposits and perform preparatory works or tunnel excavation. There has a significant effect on the economic efficiency of mining, especially in the case of deep mineral resources deposits and more challenging geological conditions.

In these conditions, a rock with very unfavourable parameters is mined. This factor applies not only to the very high value of uniaxial compressible strength, which may exceed even 200 MPa.

The high rock compactness is not only the reason for the problems with sufficient mining efficiency. Another factor is the structure of the rock. When the rock is homogenous, not layered, the cutting resistance increases drastically, and it causes enormous energy consumption and very high wear of mining tools.

The impact of the rock's macrostructure on its workability can be noticed when comparing the relationship between the compressive strength σ_c and the tensile strength σ_t. It is assumed that at $\sigma_c/\sigma_t < 10$, the rock is difficult to excavate (homogenous rock), while at $\sigma_c/\sigma_t > 15$, the rock is very easy to excavate (layered rock).

Another important factor is the content of minerals and inclusions in rocks, causing fast abrasion and wear of the mining tools, and, in the case of inclusions, e.g. sphaerosiderites, the occurrence of strong sparking during operation. Also, a more profound level of exploited mineral resources depositions influences the worsening of physicomechanical properties of rocks.

The rock mining process depends mainly on the workability of excavated rock. Rock workability is a property describing a rock's resistance to mining understood as the separation of a piece of the output from the rock mass. The measure of workability is the so-called unitary mining energy—the amount of energy required to remove a unit of a rock's volume [1–5].

The primary purpose of rock mining is to separate from the rock mass the most significant possible fragments of the output, using as little energy as possible. Rock mining is generally measured using the specific energy E_w.

$$E_w = \frac{E_u}{V} \tag{1.1}$$

where:

E_u—energy supplied to excavate the rock, J,

V—volume of the excavated rock, m^3.

Thus, the unitary energy expressed as (1.1) is the energy required to excavate a unit of a rock's volume—the lower the unitary energy, the more efficient the process. However, using unitary energy as a measure of process efficiency should be approached with caution.

During rock mining, energy breaks the rock's structure by creating cracks and craters. Assuming that the dimensions of the output grains excavated by different mining methods have the same shapes, the volume of the removed rock piece V will be proportional to its linear dimensions. Assuming that the grain shape of the output is a sphere with a diameter of d, it can be stated that:

$$E_w \sim \frac{1}{d} \tag{1.2}$$

Hence, the unitary mining energy is reversely proportional to the dimensions of the removed grains of the output. When the grain dimensions increase, the energy required to excavate a unit of the rock's volume decreases.

It can assume that the energy required to break the rock—for instance, through high-energy impacts—is proportional to 1/d, and also that the removed grains of the output have the same shape, you can write that:

$$E_u \sim d^2 \text{ and } V \sim d^3 \tag{1.3}$$

then:

$$V \sim E_u^{1.5} \tag{1.4}$$

and

$$E_w \sim E_u^{-0.5} \tag{1.5}$$

The practical interpretation of the formulas mentioned above is that a machine that hits the rock, for instance, 20 times a minute with the impact energy of 10 kJ, can remove grains of the output ten times larger than a machine with the same efficiency but hitting 2000 times per minute with the energy of 100 J. Can be explained by the fact that not whole mining energy E_u is consumed to excavate the rock, as some of it is used to create new surfaces, some of the energy will be absorbed as the kinetic energy from the mined output, as well as noise and heat energy [1, 2].

The time of preparatory galleries or tunnels implementation is directly connected with the plans of deposit development or making the tunnel available for traffic. However, based on economic analyses, it can state that the higher speed of heading mining, the lower is the unitary cost of mining one current metre of a gallery [1, 2]. The gallery drilling speed depends on the type of rock excavated and the chosen mining technique. The methods of rock mining and their working technologies are described in the next section.

1.1 Mining Methods of Hard Rock Mining

The mining methods currently used for mineral resources excavation and mainly for gallery driving can be divided into three groups: traditional method using blasting material, mechanical methods and non-mechanical methods. In the traditional method using blasting material, it is necessary to drill blast holes under the planned blasting matrix. Depending on the mining and geological conditions, you can choose one of the drilling methods [1, 2]:

- percussive drilling used for very compact rocks, with a relative slow drilling speed,
- rotary drilling used for easily workable rocks, at high drilling speed,
- rotary percussive drilling, enabling drilling in hard rock at a drilling speed slightly lower than in the case of rotary drilling.

Drilling is performed with the use of percussive, rotary or rotary-percussive drills. Percussive and rotary drills with a weight of up to 30 kg can be operated manually. In the case of heavier drills and rotary percussive drills, they are usually mounted on manipulators, installed on drill rigs on a wheeled or tracked chassis.

More often are used mechanical methods, which can be divided into conventional and non-conventional. The group of conventional mechanical methods includes:

- cutting,
- milling, including cutting.

The above-described methods are used cutting tools, mainly radial, tangential and rotary-tangential picks. These tools and the way of the work will be presented in the next chapter. Other mechanical mining methods are considered non-conventional. These include methods involving:

- rock mining based on static crumpling using symmetrical and asymmetrical disk tools,
- undercutting with asymmetrical disc tools with the tangential direction of pressure force up to rock surface.

To the group of non-mechanical methods, we can include techniques such as:

- rock mining with high-pressure water jets,
- rock mining with a high-energy gas stream from thermal burners,
- rock mining with high-frequency currents,
- rock mining using electrohydrodynamic effect,
- rock mining with swelling chemical materials.

The above-described non-mechanical methods are mainly used for breakout or crushing large blocks in opencast rock mining. They are rarely used for rock mining and will not be discussed further.

1.2   The Benefits of Mechanical Technology

The process of drilling galleries and tunnels consists of four basic operations/phases. These are mining, loading, haulage and supporting of the excavation roof.

In the case of the traditional method with the use of explosives, all individual phases of the gallery face drilling process must be carried out in series, one after the other: first, drilling the holes, then filing them with explosive material, firing and ventilation, loading the excavated output and haulage, and finally supporting of the excavation. This method takes a long time, so the gallery face drilling advance is not significant [1].

Other disadvantages of this method are: generating a shock wave to the rock mass surrounding the excavation during an explosion. This phenomenon results in cracks and rock weakening, which may lead to the need for additional reinforcement of the excavation support. Moreover, the excavation profile is uneven, not adjusted to the shape of the gallery lining, which creates inconvenient conditions for the cooperation of the lining with the rock mass. It may also affect the time and safety of the support operation.

In the underground mining industry and tunnel excavation, most of the working and preparatory galleries are drilled with mechanical methods applying mining machines such as roadheaders, continuous miners and Full-diameter miners—TBMs (Tunnel Boring Machines of Gripper and Shield type). It is mainly caused, in comparison to traditional methods using explosives, by a possibility to implement a few operations simultaneously—mostly mining, loading and haulage but in the case of TBMs also supporting the roof.

Theoretical schedules of realisation of basic mining operations are presented in Fig. 1.1. Figure 1.1a and b show the comparison of gallery face mining schedules using explosives—Fig. 1.1a for the manual drilling of the holes, Fig. 1.1b for drilling with the drill rig and loading of the output using a loader. Figure 1.1c presents the schedule for gallery drilling using roadheader, and Fig. 1.1d shows the possibility of excavation using TBMs [1].

Thanks to mechanical technology, it is possible to significantly shorten the mining time of the excavation face and increase the drilling speed, even by several times. In the case of a mining method with explosives, the drilling advance per day is usually not more significant than a few meters; for roadheaders, it reaches several or more meters and using Gripper TBMs, the advance achieves several dozen meters, even up to 70 m per day.

The next benefit is a lower weakening and disturbing rocks structure, and better fitting of the lining to the gallery profile (cross-section), which makes the lining transmit more evenly distributed loads and with much lower values.

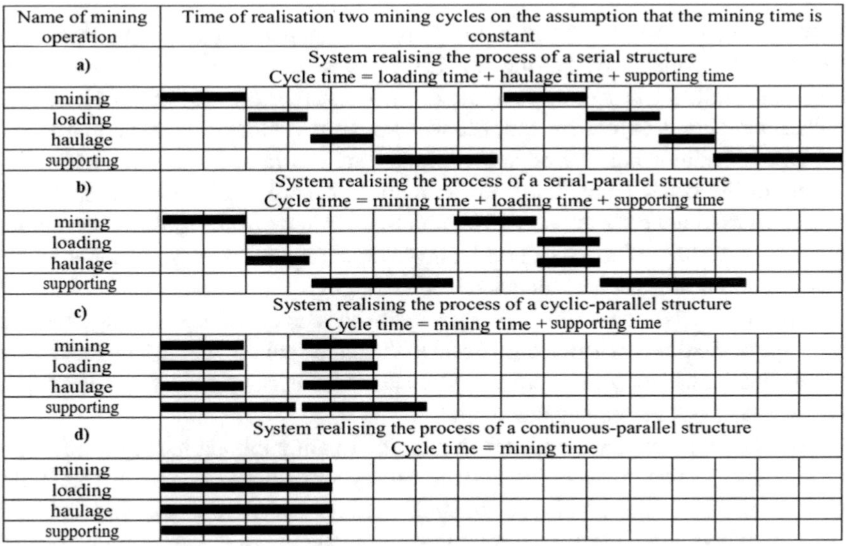

Fig. 1.1 Theoretical schedules of realisation of basic mining operations (description in the text) according to Ref. [1]

Fig. 1.2 A view of the galleries made with the use of; **a** mechanical method, **b** explosives [1]

The example of galleries executed with explosives and using mechanical methods is shown in Fig. 1.2. Compared to both galleries, the mechanical method of mining achieves even stroke and smooth wall sides, roof, and floor of the gallery. It also causes a low airflow resistance [1].

1.3 Machines and Mining Tools Used in Mechanical Methods of Hard Rock Mining

As was written above, a big part of the preparatory galleries and tunnels are drilled with mechanical methods applying mining machines such as roadheaders, continuous miners and TBMs. The examples of these machines are shown in Fig. 1.3 [1, 4]. The roadheaders (Fig. 1.3a) can be equipped with transverse or longitudinal mining heads. The types of mining heads used in roadheaders are shown in the scheme in Fig. 1.4.

Due to the better adaptability to the mining of hard rocks, nowadays, mainly transverse mining heads are used. The pointwise mining of the gallery face allows obtaining any shape of the face profile; however, it requires a longer time for whole cross-section mining. It also allows selective mining.

The continuous miners (Fig. 1.3b) are equipped with cylindrical mining heads, in the shape of the drum, with a length equal to the mined gallery width. The movement of the mining head only in a vertical plane allows the excavation of the gallery with the cross-section in a rectangular shape.

The TBMs' machines are equipped with the rotating mining head, pressed with the significant value pressure force to the gallery face. Figure 1.3c shows the TBM machine of the double shield type, which is used for tunnel mining in soft and hard rocks. It allows the gallery face mining on a whole cross-section during the one revolution, which enables to obtain the significant value of gallery face advance per day. But the gallery has a circular cross-section shape with a diameter equal to the TBMs' mining head diameter, which requires gallery artificial floor execution.

Fig. 1.3 The examples of machines used in mechanical methods of hard rock mining; **a** roadheader, **b** continuous miner, **c** double shield type TBM machine (according to Ref. [1])

Fig. 1.4 The types of mining heads used in road headers; **a** transverse, **b** longitudinal [1]

In the case of roadheaders and continuous miners, the mining process is realised by milling using mining heads equipped with cutting tools [1, 4]. Currently, there are mainly two types of cutting tools—rotary-tangential (conical) pick or radial pick, shown in Fig. 1.5. The working principles of cutting tools and problems of hard rock mining are described in more detail in Chap. 2.

The mining heads of TBMs' machines are equipped with special rotary disk tools, which mined the rock by static crumpling. In this case, mainly symmetric disk tools are used with 400–500 mm (Fig. 1.6a). They can also use asymmetric disk tools with 150–400 mm (Fig. 1.6b), which also mined the rock by static crumpling or undercutting. The working principle of disk tools and problems of hard rock mining using disk tools are described in more detail in Chap. 3.

Fig. 1.5 Types of cutting tools used in hard rock cutting or milling: **a** rotary-tangential (conical) pick, **b** radial pick (according to Ref. [4])

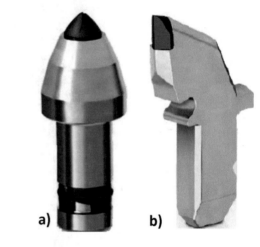

Fig. 1.6 Types of disk tools used in hard rock mining: **a** symmetric disk, **b** asymmetric disk

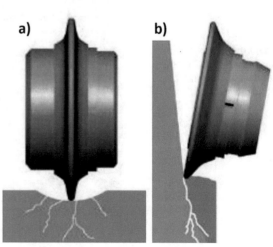

References

1. Kotwica, K., Klich, A.: Machines and equipment for roadways and tunnels mining, p. 313. Instytut Techniki Górniczej KOMAG, Gliwice (2011). (in Polish)
2. Kotwica, K., Małkowski, P.: Methods of mechanical mining of compact-rock—a comparison of efficiency and energy consumption. Energies **12**(18), 1–25 (2019). Article no. 3562
3. Kotwica, K., Prostański, D., Stopka, G.: The research and application of asymmetrical disk tools for hard rock mining. Energies **14**(7), 1–21 (2021). Article no. 1826
4. Kotwica, K.: Application of water assistance in the process of mining rock with mining tools, p. 248. Monograph. AGH UST Publisher, Kraków (2012). (in Polish)
5. Kotwica, K.: Hard rock mining – cutting or disk tools. IOP Conf. Ser.: Mater. Sci. Eng. **545**(1), 1–11 (2019). Article no. 012019

Chapter 2
Problems of Hard Rock Mining Using Cutting Tools

Abstract This chapter briefly describes the methods and mining tools used for the mechanical mining of hard rock's by cutting or milling. The first part presents the theory of rock cutting using radial and conical picks. Also, the construction and the main parameters of these tools were described. In the next part, the process of pick edge wear was presented and compared for radial and conical picks. The influence of the degree and character of the pick edge wear on the efficiency of the mining process was described. The next part presents the threats generated by cutting tools with a high degree of edgewear—mainly sparking and dustiness. The methods currently used to limit the problems described above were briefly presented. These are new materials for tools, further protection for the edges of conical picks and the use of high-pressure water spraying.

Soft and brittle rocks and rocks with naturally weakened surfaces, such as coal, chalk or limestone, are usually excavated by cutting or milling, using cutting tools—radial and rotary-tangential picks. At an appropriate pick working angle and the depth of cut, the progress of the process causes cracking in the weakest spots, with the maximum utilisation of weakened surfaces.

By acting on the rock, a cutting tool causes to varying degrees rocks to be crushed, specific rock volumes to be broken down, and elements of varying sizes to chip off. The contribution of the individual forms of rock disintegration depends on the mining type, tools geometry, operating parameters, and rock properties.

2.1 The Theory of Mechanical Mining of Rocks by Cutting

The general theory of mechanical rock mining, based on observations and the analysis of the rock mining process, was formulated. For rock disintegration, the so-called indentation zone plays the most crucial role. This theory can use both in the case of static and dynamic action on the rock. It is universal because it also considers the phenomena associated with using tools of different shapes.

K. Kotwica, *New Mining Tools and Methods for Roadheader Mining Heads*,
https://doi.org/10.1007/978-3-030-96394-1_2

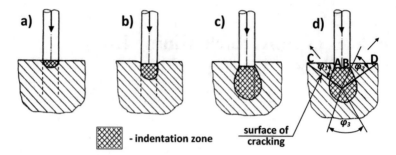

Fig. 2.1 The process of rock disintegration through pin penetration [2, 6]

The rock disintegration in this theory can be shown in the example of pressing a pin into a rock (Fig. 2.1) [1–3, 6]. The more the pressure force increases, the more the range of the indentation depth also increases (Fig. 2.1b). As the load exerted by the pin increases further, the rock shear strength along the pin's circumference (dotted line) is overcome (Fig. 2.1c), and the pin can penetrate the rock. When the pressure force achieves a sufficiently high value, the tensile stress occurring around the indentation zone causes initial internal cracks in the weakest spot (Fig. 2.1d).

These cracks are directed towards the free surface of the mined rock and cause along the pin's circumference produce from the indentation zone a wedge or an ABO cone. They make breaking off the rock elements along the OC and OD lines possible. The breakout angles φ_1 and φ_2 characterise this process, depending mainly on rock properties.

The range and shape of the indentation zone depend on type and tool geometry and the direction of the main force exerted by the tool on the rock. If the direction is tangential to the rock surface, we are dealing with cutting, if perpendicular to the rock surface—with crushing.

Mining by cutting is realized using a wedge-like or conically shaped cutting tool acting almost parallel to the rock surface. As mentioned in Chap. 1, there are mainly radial or conical picks. Figure 2.2 presents the scheme of radial pick with the main geometric parameters [1–3, 6]. The tool has a holder for mounting on the cutting head and the working part, which does the cutting, featuring a cemented carbide insert. The tool has the following angles: α-clearance angle, β-pick edge angle and γ-rake angle, the sum of which is $\alpha + \beta + \gamma = 90$.

The α plane is called the clearance surface, while the γ plane, with which the tool attacks the cut rock and throws the material out of the groove, is called the rake surface. The β planes facing the sidewalls of the groove are called side rake planes. The cutting process generates a cut with the depth g. For the machine to cut the rock instead of crushing it, the clearance angle must be $\alpha > 0$.

Depending on the physicochemical properties of the excavated rock and the pick shape, to cut the rock to the depth g, the force P must be applied to the pick, opposite to the force coming from rock mining resistance. The force P can be distributed into three constituent forces exerted in three perpendicular directions (Fig. 2.3): the force

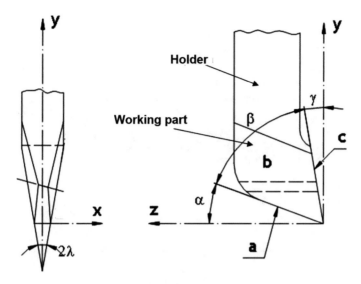

Fig. 2.2 The scheme of radial pick with the main geometric parameters [3, 6]

Fig. 2.3 Diagram of the forces acting in the general case on the radial pick [3, 6]

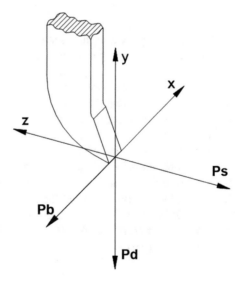

exerted parallel to the mined rock surface—cutting force *Ps*, which plays the crucial role in the cutting process, the force exerted perpendicular to the excavated rock surface—the pressure force *Pd*, its value determines the value of the cutting depth *g*, and the force that is perpendicular to them—the lateral force *Pb* [2, 3, 6].

Since rocks are more brittle than plastic, in the cutting process, it produces, instead of continuous chips, as with metal cutting, spoil consisting of various sizes of grains and dust. The pick's edge movement creates a groove whose shape, especially the

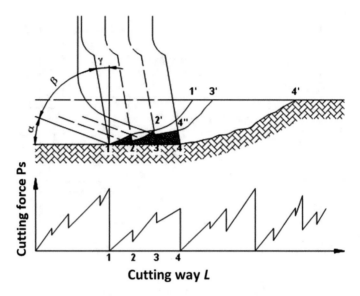

Fig. 2.4 Scheme of changes in the cutting force during mining using radial pick according to Refs. [3, 6]

bottom, is similar to that of the pick (the width b and the shape of the cutting edge). The scheme of the cutting process of rock using radial pick shows Fig. 2.4 [1–3, 6].

After a more significant grain breaks off, the pick drives with its cutting edge into the frontal, inclined wall of the fissure, cutting a certain amount of the rock (the dotted area). The cutting force Ps value increases gradually. This situation makes it possible to arise a complex state of stress in the rock, causing tensile and shear stresses to be more efficient in rock disintegration than compression stresses.

Once the contact area of the pick edge and cutting force Ps value increase to a specific value, the cohesion of the rock is overcome. This phenomenon leads to the breakage of the rock chips and throwing them out of the cut groove, with the cutting force Ps value decreasing. When the pick moves further, the rock is again crushed, with the cutting force and the contact area between the pick and the rock increasing until the next larger chip breaks off. The time between such break-offs of larger chips is called the cutting cycle.

2.2 Pick Edge Wear and Its Influence on the Efficiency of the Cutting Process

Pick edge wear has a significant influence on the efficiency of the cutting process. This phenomenon depends on rock properties and their disintegration during the mining process. The wear has dynamic nature, its shape and intensity varying as a function of time. Wear occurs when the pick loses some of its volume (through abrasion or micro-cutting) due to mechanical fatigue (resulting from dynamic loads) or thermal fatigue [1, 3]. This phenomenon leads to a change in the pick's geometry and its contact area with the rock.

Figure 2.5 shows the usual wear pattern of V-shaped radial picks used in rock cutting [3, 6]. The first sign of the edge wear is increasing radius r of the pick's rounded edge (point-1). In the next stage, the contact of the clearance surface with the bottom of the cut groove gradually flattens and increases. The width S_p of this flattened area is considered a measure of pick edge wear (point-2 and 3). This flattening is inclined concerning the tool's actual movement direction W at a slight angle of τ. Pick wear can also arise due to splintering and micro-splintering on the tool's rake plane and side planes. For the efficient cutting process with a blunted pick, the cutting depth g must be larger than the pick's rounded edge radius r.

During the rock cutting with the blunted pick, the rock indentation zone occurs both in front of and underneath the pick (Fig. 2.6). The reason is the larger contact area of the blunted pick edge with the mined rock. The output (area 1) is forced sideways or upwards, but due to the pressure from the tool edge face, it carries a small layer (2) of crushed and pressed rock residue. This residue is temporary and successively replaced [3, 5, 6].

The more significant the pick edge's wear and the greater its contact surface with the bottom of the cut, the greater the crumpled rock zone under the clearance surface (3). This zone is directly responsible for the production of dust.

The degree of the tool's wear has a significant influence on the efficiency of the mining process. Studies performed when excavating coal using longwall shearers have shown that the blunting of radial picks edge has a significant impact on the electric power consumption during mining, particular specific power consumption, concerning the volume of the excavated rock [1, 3, 6].

Fig. 2.5 Successive stages of wear in a V-shaped radial pick [3, 6]

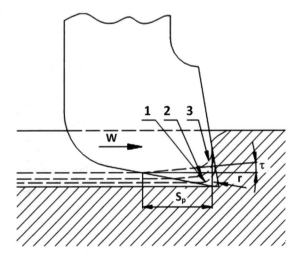

Fig. 2.6 Cutting with a
blunted V-shaped radial pick
[3, 6]

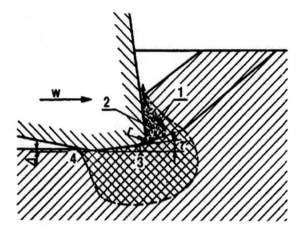

Figure 2.7 shows the relationship between the power consumption N and the specific power consumption coefficient K_u as a function of the longwall shearer's feed speed v_p for four values of radial picks edge blunting S_p in millimetres (see Fig. 2.5).

Comparing the charts shows that the increase of the width S_p of the flattened area of picks edge wear will affect a lot for power consumption. This fact stays in

Fig. 2.7 The impact of the
radial-pick blunting S_p on
the power consumption N
and the specific power
consumption coefficient K_u
as a function of the longwall
shearer's advance rate
according to Ref. [3]

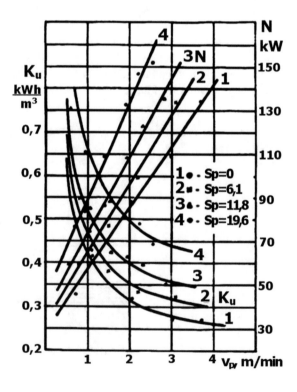

connection with the rise of cutting depth g, which increases with the growth of the shearer's feed speed v_p. Compared new pick and the pick with the width S_p of the blunting area equal 19.6 mm, the power consumption K_u is more than double. The bigger blunted surface increases edge heating and a sparking tendency, and according to Fig. 2.5, the amount of dust production [3].

This phenomenon increases with the growth of the compressive strength of the excavated rock. Due to this reason, currently, it is more common to use conically shaped picks that can rotate in the holder (conical picks).

The construction of a conical pick is distinguished by a cylindrical handle and a conically shaped working part, ending with a sintered carbide edge. The pick edge has a revolving body, most often cylindrical and conical, and is mounted in the working part body's bore by soldering.

The conical shape of the pick influences the spatial form of the crushing area, as unlike in traditional picks, by design, the pick comes into point contact with the rock, not into linear contact. Consequently, the crushing area has a smaller range, and the unit loads on the rock are more significant, which is essential for compact-rock mining.

The geometry of these picks is closely associated with the mining process. The pick angle is determined similarly to that of radial picks. The position of the tool concerning the cutting line depends on the setting angle and the side deflection angle. The geometry of the conical pick in a static and movable reference system shows Fig. 2.8 [1, 3].

The tangential rotary picks in the mining heads are mounted directly in the unique holders or with hardened ex-changeable sleeves. The tools are set up in the holders at the given angles to maintain their proper operation. These angles and the way of assembly are shown in the scheme in Fig. 2.8.

The constant factors of a proper operation of conicalpicks are [3]:

- their right setting (selection of the optimum cutting angles),
- the way they are mounted,
- the bearing in the tool holders.

When the conditions for mounting the rotary tangential pick in the holder are correct, it allows free and regular turnover of the pick. It causes symmetrical and very even wear of the pick and increases tool life. Such ideal wear of the cutting tool—conical pick, is shown in Fig. 2.9 [3].

When the rotary-tangential pick is heavily blunted, the reaction force of mined rock is powerful, and as the edge wear progresses, it leads that pressure force Pd increases substantially, while the cutting force Ps increases only slightly. In the case of a conical pick, instead of the pick wear width S_p, on the side of the tool face, we can use the radius d_s of the rounded edge due to blunting (Fig. 2.10). Assuming that edge wear occurs uniformly, the wear causes only pick shortening by the value h_n, without any changes in the pick's edge angle [3].

In practice, the wear process of these picks is much more complex and varied [3]. The significant influence on the wear has mechanical damages (splintering and insert break-off) or damages due to mechanical or thermal fatigue (usually cracking).

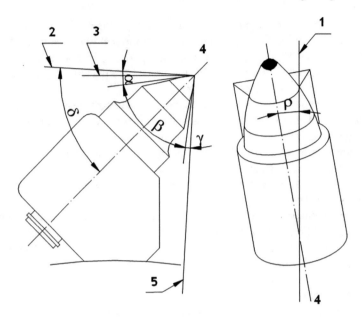

Fig. 2.8 The geometry of the conical pick in a static and movable reference system according to Refs. [1, 3]: 1—tangent to the cutting surface by the cutting head, 2—tangent to the cutting line, 3—cutting line, 4—pick axis, 5—radius of the cutting head, α—clearance angle, β—edge angle, γ—rake angle, ρ—side deflection angle, δ—setting angle

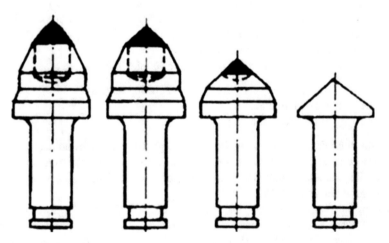

Fig. 2.9 Phases of an "ideal" wear process for a conical pick [3]

Fig. 2.10 Wear pattern of a
conical pick [3]

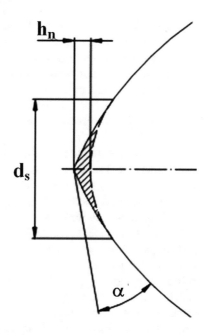

The above-described mounting way of bearing generates the occurrence of dry
friction that often leads, in connection with low torque value, to seize tool handles in
the holders, deteriorate their rotations and cause their premature asymmetric wearing.
This situation applies to more than 50% of tools used in the mining industry. Pick
rotation plays an essential role in pick wear. In compact rocks stopping or significantly
reducing the number of revolutions of the pick results in abrupt and asymmetrical
pick wear [1, 3].

Figure 2.11 presents very irregular wear of cone pick edge in the absence of
rotation of the tool in the holder. This result was obtained on a laboratory stand after
2 min of the cutting process in incorrect conditions. Such an edge wear pattern leads

Fig. 2.11 An influence of a
conical pick rotation decay
on the edge wear, a
laboratory tests result

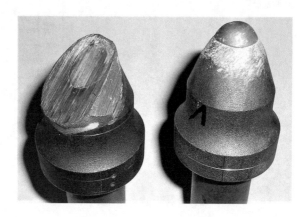

to increased pick loads, especially the pressure force value Pd, the growth of dust generation and cutting head vibration.

Tool wear is inevitably accompanied by a more extensive fragmentation of output, creating more dust. It is several times larger than in a sharp pick edge. The laboratory tests have shown that even an insignificant increase in tool wear is enough to cause increased dustiness by 30–70% [3, 4]. Figure 2.12 shows the impact of the pick wear on weight loss on the concentration of respirable dust when rock mining using conical picks.

Mechanical cutting and milling methods are not appropriate for complex and very compact rocks, especially those containing silicon and volcanic-rock inclusions. Such rock massive requires the use of large cutting forces. The gallery excavation using roadheader in rocks with a uniaxial compression strength of more than 150 MPa results in extensive cutting tool wear, in extreme cases reaching five picks per 1 m³ of cut material. As a result, the driving efficiency decreases substantially, and the costs rise sharply. It is considered cost-effective to use tools with a wear level of about 0.2 picks per 1 m³ of cut material [3]. As a result, the driving efficiency decreases substantially, and the costs rise sharply.

The disadvantage of mining rocks by milling or cutting, using cutting tools, is the generation of dust and limitation of the use of this method, associated with the upper limit of rock uniaxial compressive strength to or high compactness of these rocks. In connection with the high abrasiveness of such rocks, this results in excessive wear of cutting picks and a decrease in drilling speed. The significant wear of the tool edge also increases the dustiness and the risk of explosion of mine gases, mainly methane, resulting from frictional sparking. These threats, caused by excessive wear, are presented in Fig. 2.13. This problem occurs mainly with radial picks, but even using a conical pick, in the case of compact rock, reduces this problem only to a small extent [1, 3].

Fig. 2.12 Respirable dust concentration as a function of pick weight loss according to Ref. [3]

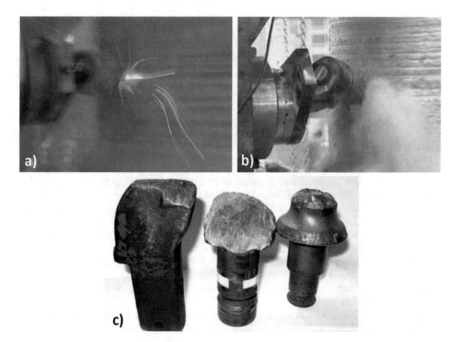

Fig. 2.13 A view of the threats generated during cutting of a rock sample with cutting tools: **a** frictional sparking, **b** dust generation, **c** pick edge wear according to Refs. [2, 3]

2.2.1 Possibilities of Increasing the Durability of Conical Picks

When mining hardly mineable rocks, the increased wear of tools and the generated dust hazards and gas explosion often result in rock mining being economical unprofessionally (low efficiency and high energy consumption of the mining process, increased wear of tools). Work is underway to develop more durable cutting tools; however, even new-generation rotary-tangential pick cannot always provide the required mining parameters [1, 3]. Actions taken in this direction are presented below.

- New designs of tangential-rotary picks have been developed, equipped with unique rings securing the carbide insert against breaking.
- New sintered carbide inserts are used—mushroom-shaped or hat-shaped, which protect the working part of the pick against abrasion.
- Carbides are made of materials with a fine-grained structure, increasing their durability.

But even very high-quality cutting tools, especially conical picks, are exposed to increased wear under incorrect operating conditions related to the loss of rotation of the picks in the holders. These inconveniences prompted the search for alternative cutting tools to be mounted on the mining heads of roadheaders in the place of conical picks. The effects of these works are presented in the following chapters.

Whereas for assurance of the required working conditions in holders for conical picks, few ways of high-pressure water assistance have been developed and tested. These are presented in Chap. 5.

References

1. Kotwica, K., Klich, A.: Machines and equipment for roadways and tunnels mining, pp. 313. Instytut Techniki Górniczej KOMAG, Gliwice (2011). (in polish)
2. Kotwica, K., Małkowski, P.: Methods of mechanical mining of compact-rock—a comparison of efficiency and energy consumption. Energies **12**(18), 1–25 (2019). Article no. 3562
3. Kotwica, K.: Application of water assistance in the process of mining rock with mining tools, p. 248. Monograph. AGH UST Publisher. Kraków (2012). (in polish)
4. Kotwica, K.: Effect of selected working conditions of cutting picks on their wear during the mining of hard rocks. Mech. Control **29**(3), 110–118 (2010)
5. Kotwica, K.: Hard rock mining—cutting or disk tools. IOP Conf. Ser.: Mater. Sci. Eng. 545(1), 1–11 (2019). Article no. 012019
6. Opolski, T.: Urabianie mechaniczne i fizykalne skał. Wydawnictwo Śląsk, Katowice, Poland (1982). (in Polish)

Chapter 3
The Hard Rock Mining Using Disk Tools

Abstract This chapter briefly describes the methods and mining disk tools used for the mechanical mining of hard rocks. The first part presented the theory of rock mining using symmetrical disk tools. The mining method of quasi-static rock crushing using a smooth disk tool was described. The second part includes the description of symmetrical disk tools used in mining heads of TBMs machines—with smooth edges, with edges armed with carbide insert and with toothed edges, in one and double edges versions. The construction, the main parameters, and essential advantages and disadvantage of symmetrical disk tools mined the rock by static crushing were presented. The next part presents the principles of the unconventional method of rock mining using asymmetrical disk tools called the back incision or undercutting method. The machine solution used the method, and the obtained effects were described. The advantages and drawbacks of this mining method, using big diameter asymmetrical disk tools, were presented.

Another of the most commonly used methods in mechanical mining of hard rock is mining through static crushing, carried out with the help of wheel tools called disks. The disk tools used for this mining method can be symmetrical or asymmetric (see Fig. 1.6). In this method, the disk edge is pushed into a rock massive with the normal force perpendicular to its surface.

The technique of rock mining based on static crumpling using symmetrical and asymmetrical disk tools is shown in Fig. 3.1 [1, 8]. The disk tool has a diameter D and a cutting-edge angle equal β-in the case of asymmetrical or 2β-in the case of symmetrical disk tool. In this method, the disk's edge, with a V-like, symmetrical or asymmetrical shape of its cross-section in the plane normal to its edge, is pushed into the rock with Pd normal force perpendicular to its surface. As a result of this force, the rock's compressive strength is locally exceeded, with the disk tool driving into it to the penetration depth h. Besides, the tangential or rolling force Ps is applied to the disk handle, causing the tool to move along the rock's surface. Due to rolling force Ps and normal force Pd, the disk puts the rock under pressure coming from the resultant force P along with a circumference corresponding to the angle φ_d. The disk tool is mounted in a rotary holder, enabling it to rotate on its axis with the angular speed ω.

© The Author(s), under exclusive license to Springer Nature Switzerland AG 2022   23
K. Kotwica, *New Mining Tools and Methods for Roadheader Mining Heads*,
https://doi.org/10.1007/978-3-030-96394-1_3

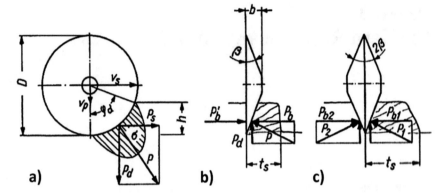

Fig. 3.1 The method of rock mining based on static crumpling: **a** general scheme, **b** the way of mining using asymmetrical disk tool, **c** the way of mining using symmetrical disk tool (description in the text) [1, 8]

3.1   The Principle of Mining with a Smooth, Symmetrical Disk Tool

The physical model describing the process of static symmetrical disk tool penetration into the rock is based, as in the case of other mechanical mining methods, e.g. cutting, on the crushing zone hypothesis. However, a different shape of the edge tooltip should be taken into account.

Smooth, symmetrical disk tools can be classified as complex rolling profiles, where the disk tool profile is a radial composition of a specific set of curvilinear wedge profiles [1, 8]. It should be noted that the disc tool profile embedded in the rock body to a certain depth (limited by the height of the disc edge) generates the occurrence of an area of damage identical to that resulting from penetration into the rock mass of the curvilinear wedge. Concerning the theory of penetration of the wedge tool into the rock semi-space known in the literature, the course of the qualitative phenomena observed in the process of the destruction of the spatial structure of compact rocks with a quasi-static disc tool can be described by distinguishing its following phases (Fig. 3.2) [1, 8]:

- In the initial phase (I) of the first penetration of the disc tool edge into the rock semi-space (with the linear rolling speed equal to zero), due to the concentration of the point stresses, surface crushing of the rock occurs, resulting in the formation of the so-called a zone of a compacted product of crushing. The development of the compacted zone takes place as long as micro-cracks do not appear at the boundary surface of the contact of the crushing product with the rock mass. In hard rocks, the first micro-cracks appear at the top of the compacted zone or directly in its vicinity, and their extent is of the order of a few millimetres.
- In the following, the so-called developed phase (II) of the vertical penetration of the disc tool edge, the micro-cracks propagate radially into the rock half-space.

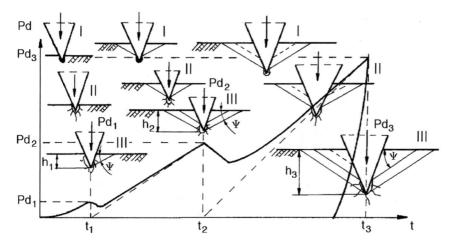

Fig. 3.2 Destruction phases accompanying the process of vertical penetration of the disk tool into the rock massive: Pd—normal force, t_s—mining spacing, Ψ—breakout angle, h—penetration depth [1, 8]

Micro-cracks near the top of the compacted zone propagate more profound into the rock mass faster than the others and lead to the formation of the so-called major cracks (one or more at once). The extent of the major cracks may be in the order of several millimetres and several centimetres in some cases.

- In the final phase (III), referred to as shearing off the side rock fragments, the rock fragments are broken off on both sides of the active part of the disc edge. Shearing of the side fragments of the excavated material takes place along the planes of the least resistance, inclined at a specific angle Ψ called breakout angle, to the free surface of the excavated rock mass. For a particular type of rock, the breakout angle Ψ keeps an approximately constant value.

At the end of the shearing process of the side rock fragments, the system: disk tool—rock is wholly or partially relieved. A further increase in the penetration of the disc tool edge in the rock half-space causes the reappearance of the I, II and III phases of the destruction of the rock mass, with the fact that the area of the compacted product of crushing and the side-cut surfaces in the second sequence of the disk tool load condition (and in each subsequent one) are correspondingly larger.

The graphic interpretation of the boundary conditions for the pressures in the contact zone of the disc tool with the rock half-space, after obtaining the value of the force F, respectively, exceeding the shear strength of the rock sample and the uniaxial compression strength of the rock sample, is shown in Fig. 3.3 [1, 8]. Figure 3.3a presents the interpretation of the boundary conditions of the shear loads in the elementary area of the rock massive above the $x_1 - x_2$ plane, which is responsible for the breakage of the rock fragments sideways at the breakout angle Ψ.

Whereas Fig. 3.3b shows the interpretation of the boundary conditions of the compressive loads in the elementary area of the rock medium below the $x_1 - x_2$

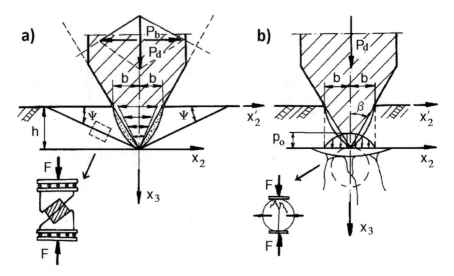

Fig. 3.3 The graphic interpretation of the boundary conditions for the pressures in the contact zone of the disc tool with the rock half-space [1, 8]: **a** the boundary conditions of the shear loads, **b** the boundary conditions of the compressive loads b—geometric dimension of the contact surface of the disc tool with the rock, F—the force destroying the rock sample (other markings as in Fig. 3.1 and 3.2)

plane, which is responsible for the formation of the crush zone and the formation of the major cracks.

3.2   The Advantages and Disadvantages of Hard Rock Mining Using Symmetrical Disk Tool

The typical construction of the symmetrical disk tool consists of shaft, seal, hub, split ring and the replaceable cutting ring and is shown in Fig. 3.4. The symmetrical

Fig. 3.4 The typical construction of the symmetrical disk tool

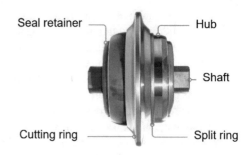

disk tool's diameter is usually around 400 mm, but there are also disk tools, which diameters can be even up to 500 mm [3, 6, 7].

There are three types of symmetric disk tools: with smooth edges, with edges armed with carbide insert and with toothed edges, in single and double or even triple edges versions. However, the single or double edges versions are most often used for mechanical mining of galleries and tunnels.

The type of disk tool is selected depending on the uniaxial compressive strength of the rock σ_c [3, 7]. For rocks with compressive strength σ_c in the range of 120–180 $_{MPa}$, disks with smooth edges are used, for compressive strength σ_c in the range from 150 to 200 $_{MPa}$—disks with toothed edges, and for compressive strength σ_c above 200 MPa—disks with edges armed with carbide insert. However, the number of cutting edges of the disk tool depends on the tool position on the mining head of the TBM machine. The greater the load acting on the tool (e.g. closer to the axis of the head or its circumference), the more frequently double-edged or even triple-edged tools are used [3, 6]. The examples of single and double edges disk tools with smooth edges and edges armed with a carbide insert are shown in Fig. 3.5.

TBM disk tools are mounted on the TBM mining head in a specific pattern and cover the gallery face. They are moving on the circle trajectories, with the spacing distance t_s between the disk tools. After one turnover of the mining head, the gallery face is excavated over the whole cross-section to the depth of the disk tool penetration. This rock mining method ensures a huge advance per day, up to 70 m or even more [5, 7].

Fig. 3.5 The examples of single and double edges disk tools with smooth edges and edges armed with a carbide insert [3, 6]

The essential advantage of mining with symmetrical disk tools is that it radically reduces the role of friction forces by using the rotary movement of the disk tool. Moreover, due to rotary motion, each section of the disk's edge remains in contact with the rock for a short period, during which it does the work required to achieve the target needed depth of cut. This results in more minor energy losses and better heat dissipation conditions on the disc's edge, reducing the hazardous of mining, such as sparking due to local increases in temperature or dust generation and extending the tool's service life.

A smooth single edge disk tool can work without visible wear even up to a few kilometres of roadway or tunnel advance. The disk tool with an edge armed with a carbide insert is sometimes even twice as durable. In comparison, the cutting tool life is not more than several cubic meters of output per piece.

The significant disadvantage of mining by static crushing is the need to ensure a high-value normal force for the pressing of the disk tool. Destruction of the stone rock structure occurs due to exceeding its uniaxial compressive strength σ_c, which value is, in the case of the rock, the biggest. According to the available data, the value of the normal force per single disk tool, enabling its penetration in the rock to a sufficient depth, can be as high as 300 kN. In the case of many disk tools on the mining head, give a considerable value of the total pressure force, which depends on the diameter of the mining head and disk tools spacing. The value of total pressure force can achieve for big diameter TBM machines up to 25,000 kN. The mining machine must transfer the reactions coming from the mining head. To stabilize the TBM machine during operation, a several times greater side force value is required to expand the machine in the excavation between the side walls. These factors generate the large mass (up to 3500 Mg) and dimensions of the machine (the length of the TBM reaching over 400 m). Most of the gripper TBM machines is designed and manufactured as a single prototype production, which increases the cost of their implementation [5, 7].

This solution results in the limited use of this type of machinery to drive excavations with a very large runout (in the dependence of TBM diameter, the length of gallery or tunnel must be no shorter than 1.5 up to 5 km). In this case, the huge costs related to their production and implementation in the mining excavation can be paid back [3, 7]. The view of one of the gripper TBM machines with a diameter of 6.5 m and the mining head equipped with symmetrical disk tools is shown in Fig. 3.6.

The attempts were made to use high-pressure water jets for mining process assistance to reduce the value of the forces acting on the disk tools and thus to limit the dimension, weight and price of the TBM machine. The attempts gave positive results, but they were not applied on a large scale.

Fig. 3.6 A view of one of the gripper TBM machines with a diameter of 6.5 m and the mining head equipped with symmetrical disk tools according to Ref. [3]

3.3 Hard Rock Mining Using the Undercutting Method and Asymmetrical Disk Tool

The asymmetrical disk tools, except for their use in a similar mining method by static crushing, have been used as mining tools in mining heads of longwall shearers to increase large grain output. The industrial tests demonstrated the usefulness of such equipment for obtaining higher graining, but increased dynamics of mining with the mining heads equipped with asymmetrical disk tools were currently not commonly used.

Asymmetrical disk tools are applied in mechanical mining not only as crushing tools but also as chipping ones, in the rock mining through breaking with an asymmetrical disk tool, typical for rocks much smaller shear strength τ value than uniaxial compressive strength σ_c was used. The smaller value of resistance against stretching enables the application of the forces of smaller values to the disk tool. This mining technique is called the back incision or undercutting method. Figure 3.7 presents a diagram of rock mining with the undercutting method [2, 4, 5, 9].

The principle of an undercutting method is mining a rock by cutting it off towards free space. A disk tool affects the rock tangentially to the surface of the mined rock massive, similarly as in the case of a cutting tool. The difference is that the tool performs the rolling movement, which efficiently eliminates sliding friction in favour of rolling friction of much less value. Implementing the undercutting method using asymmetrical disk tools significantly lowers energy consumption and pressure force and increases the output grains size.

The reduction of disk tools load allows constructing a mining machine of respectively lower energy parameters and dimensions than in the case of classical TBM machines using crushing disk tools, operating perpendicularly towards the surface of the mined rock massive.

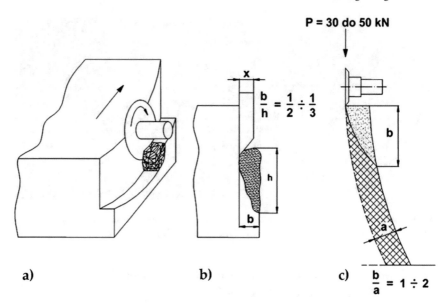

Fig. 3.7 Diagram of the undercutting mining method principle: **a** general scheme, **b** parameters of the output piece chipped from the solid rock, **c** parameters of the mining process [9]

However, in the case of the undercutting method, the side forces on disc tools edges have significant and highly variable values. It results from the proper transfer of reactions on disk tools holders and their bearing. In the scheme, in Fig. 3.8, this phenomenon is explained. In Fig. 3.8a, the rock is not chipped off yet, the disk tool cuts into the rock body, and in Fig. 3.8b, the rock chip is already broken out. The disk tool cuts a new mining groove in the rock and is initially pushed away from the rock massive. The lateral force vector Pb changes its sense to the opposite [2, 9].

The diagram and view of a prototype of the machine called Hinterschneiden machine, which uses the undercutting method developed and manufactured by the Wirth Company, is presented in Fig. 3.9. The machine (Fig. 3.9b) features four

Fig. 3.8 Diagram of the forces affecting the edge of asymmetrical disk tool in the undercutting mining method [9]

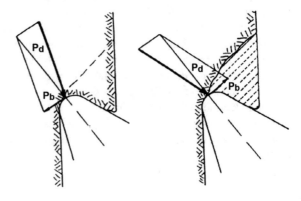

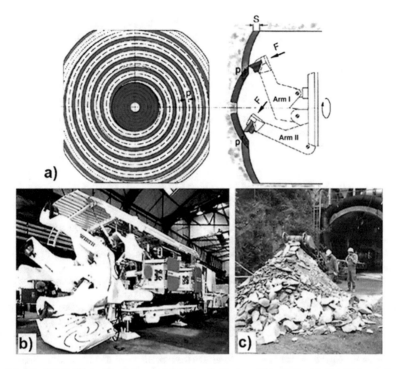

Fig. 3.9 The Wirth Hinterschneiden machine uses the undercutting method: **a** the diagram of work principle, **b** view of a prototype of the machine, **c** the view of big dimension grains of output after the mining process according to Ref. [9]

independently controlled, swinging and rotating arms, equipped with asymmetrical disk tools that can mine in any way the surface of the gallery face. The machine was applied for drilling excavation of dimensions 4 m × 4 m in Carboniferous sandstones, about uniaxial compressive strength 120 ÷ 140 MPa and quartz content up to 74% [2, 4, 9].

Each arm of the machine mining head was armed with a disk tool 450 mm in diameter. Swinging and rotary movement of the arms caused that the movement trajectories of individual disks created spirals (Fig. 3.9a). With the machine's installed driving power of 70 kW, a penetration depth h of 6 mm per rotation and disk tool was achieved for the predefined advance depth of 120 mm. This solution allowed achieving unit mining efficiency of 3 m³/h per disc. For comparison, the traditional symmetrical disk tool, using the TBMs machine, provided a mining efficiency of only about 1 ÷ 2 m³/h. The solution also had a significant advantage of there being no visible and measurable wear signs on the discs and obtaining big dimension grains of output (even over 200 mm—Fig. 3.9c) [3, 9].

The undercutting method used in the presented machine showed the whole usability of the suggested solution. Auspicious results were obtained, especially significantly lower unitary energy consumption than in the case of TBMs miners and a few times lower than for roadheaders. The only drawback is a highly complex method of the steering system for individual arms and a very high value of reaction forces.

It was the reason that the idea of the undercutting method for hard rock mining was implemented in the Department of Mining, Dressing and Transport Machines, AGH University of Science and Technology, Cracow, to elaborate an innovative construction of a mining head for roadheaders, equipped with mini asymmetrical disk tools. The course of implemented works is presented in Chap. 6.

References

1. Korzeń, Z., Lewicki, M.: Obciążenia zewnętrzne narzędzi dyskowych w procesie urabiania skał twardych Arch. Min. Sci. **35**, 279–304 (1990). (in Polish)
2. Kotwica, K., Gospodarczyk, P., Stopka, G.: A new generation mining head with disc tool of complex trajectory. Arch. Min. Sci. **58**(4), 985–1006 (2013)
3. Kotwica, K., Klich, A.: Machines and Equipment for Roadways and Tunnels Mining, p. 313. Instytut Techniki Górniczej KOMAG, Gliwice (2011). (in Polish)
4. Kotwica, K., Małkowski, P.: Methods of mechanical mining of compact-rock—a comparison of efficiency and energy consumption. Energies 12(18), 1–25 (2019). Article no. 3562
5. Kotwica, K., Prostański, D., Stopka, G.: The research and application of asymmetrical disk tools for hard rock mining. Energies 14(7), 1–21 (2021). Article no. 1826
6. Kotwica K.: Application of Water Assistance in the Process of Mining Rock with Mining Tools, p. 248. Monograph. AGH UST Publisher, Kraków (2012). (in Polish)
7. Kotwica, K.: Hard rock mining – cutting or disk tools. IOP Conf. Ser.: Mater. Sci. Eng. **545**(1), 1–11 (2019). Article no. 012019
8. Kotwica, K.: Urabianie hydromechaniczne skał zwięzłych narzędziami dyskowymi. Dissertation, p. 276. AGH University of Science and Technology, Cracow (1996). (in Polish)
9. Weber, W.: Drążenie chodników o różnych przekrojach przy pomocy techniki tylnego wycinania. Maszyna firmy Wirth-HDRK "Continuous Mining Machine". Sympozjum "Drążenie chodników w górnictwie węgla kamiennego". Siemianowice Śl (1994). (in Polish)

Chapter 4
The Crown Pick as an Alternative for the Conical Pick

Abstract This chapter describes the new solution of a crown pick as an alternative for the conical pick. The construction and work principle of a crown pick was described. In the second part, the results of the laboratory tests using new crown tools and standard conical pick were presented and compared. The advantages of crown pick were described. Based on the laboratory stand tests results, the next part describes the choice of the most convenient crown pick solution for the subsequent trials—lowest load and the tool's wear, the highest number of tool rotations. The fourth part includes the analysis description connected with the new solution of crown pick directly on the roadheaders mining head. The results of the works allowed the manufacturing of two sets of crown picks for field and industrial tests. The conducted field tests confirmed the possibility of using a new crown pick as an alternative for conical picks. The wear of the tool was slight. The only disadvantage was the problem with the steering system not allowing for gentle and slow control of mining head movements. The results of industrial trials also confirmed this inconvenience.

As mentioned earlier, the tangential-rotary (conical) picks require regular rotation in the holders to ensure even top pick wear. This rotation is ensured by the appropriate value of the torque generated by the lateral forces acting on the pick top and the working part of the pick. The value of these lateral forces is determined by the appropriate setting of the conical picks on the mining head—the value of the side deflection angle ρ and the value of the twist angle κ of the conical pick. But even the correct setting of the tool on the mining head cannot ensure its rotation in the holder in many cases. This case is due, for example, to the high value of friction coefficient between the pick shaft and the inner surface of the holder. The tool rotation in the holder can be facilitated by reducing the value of the friction coefficient or by increasing the value of the torque generated by side forces. The first aspect will be presented in Chap. 5, while this chapter will describe the solution of an innovative cutting tool that allows obtaining much higher torque values acting on the working part of the cutting pick.

© The Author(s), under exclusive license to Springer Nature Switzerland AG 2022
K. Kotwica, *New Mining Tools and Methods for Roadheader Mining Heads*,
https://doi.org/10.1007/978-3-030-96394-1_4

4.1   The Construction and Principle of Work of the New Solution of the Crown Pick

The new solution of a unique cutting tool, which can replace traditional conical picks on the mining head of the roadheaders, was developed in AGH UST. This tool was adapted to attach standard holders of conical pick, commonly used for hard rock cutting with the compressive strength σ_c about 100–120 MPa. The difference was in the construction of the working part of the pick. The scheme of the new cutting tool, compared to the standard conical pick, is shown in Fig. 4.1. The handle part of the tool is the same for both picks—it has a diameter of about 30 mm, and the length in the case of standard conical pick equal 80 mm (Fig. 4.1a) and for new pick 75 mm (Fig. 4.1b—pos. 1) [1, 4].

The standard conical pick is equipped with a single carbide insert in the cone shape with a diameter of 22 mm, mounted in the axis of the tool. The working part of the new tool was developed as a bell or crown-shaped cutting segment (Fig. 4.1b—pos. 2). Hence the name of the new tool—the crown pick. The working part is armed at

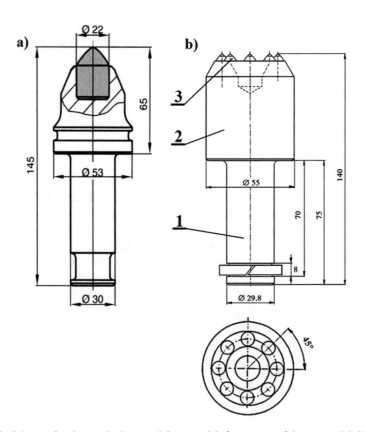

Fig. 4.1 Scheme of: **a** the standard tangential-rotary pick, **b** new type of the crown pick [1, 3]

the circumference with eight cone-shaped sintered carbide inserts with a diameter of about 8 mm (Fig. 4.1b—pos. 3).

Instead of traditional cutting using standard tangential rotary picks, such a design solution should allow chipping rock fragments due to point pressures exerted by individual sintered carbide inserts.

The additional advantage of the new pick solution is the non-uniform load of individual sintered carbide inserts, which should cause an increase in the value of torque working on the tool. Thus, it should also allow for increasing the tool rotary speed in the holder at a minimal side deflection angle of the cutting tool.

The new solution of the crown tool has been submitted to the Polish Patent Office and is covered by patent protection No. PL 182135 [2].

After manufacturing the new tool prototype, preliminary tests at a unique laboratory stand were carried out. The tests were performed on the special laboratory test stand for single tools testing, which was developed to create similar to natural working conditions for single mining tools.

4.2 The Comparative Laboratory Stands Tests of a New Type of Crown Pick and Conical Pick

The comparative tests of a new crown pick and conical pick on the special test stand were scheduled. The laboratory stand used for the tests is shown in Fig. 4.2. The laboratory stand has a unique construction consisting of a frame 1, and a traverse 2 is moved vertically using screw drives. On the traverse is mounted the sliding support 3 with a mining tool holder 4 of L shape. The support can move along the traverse using the hydraulic cylinder. In the mining tool holder, the measuring head, mounted on the bottom part of the long arm of the tool-holder, enables an independent measurement of cutting force components: tangential force Ps, pressure force Pd and side force Pb.

For the tests with a new type of crown pick and conical pick shown in Fig. 4.1, the standard holder used in roadheaders mining heads, in which a cutting tool is directly mounted, was applied.

The pick holder was welded to the special plate. Through movable mounting with the measuring head, this plate can be rotated in relation to the axis of the measuring head. This solution allows the location of the cutting pick in the holder to cut in the layout close to that occurring on the cylindrical part of roadheaders transverse mining heads. The cutting pick is mounted in the holder at a setting angle $\delta = 45°$. The side deflection angle value ρ can change in 3° increments, in the range of $\rho = 0°$ up to 45°.

In the axis of the test stand, the rotary table rotated by a hydraulic drive 6 is mounted. On the table, the special artificial concrete, or natural stone sample 5 in the shape of the ring, is fixed. The test tool mined the ring sample on the side surface, from the top to down, with set cutting depth, cutting speed, and spacing.

Fig. 4.2 A view of the
laboratory stand for single
mining tools description in
the text

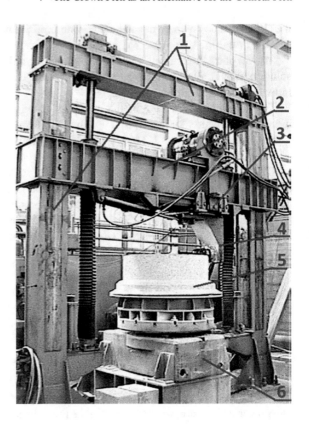

The comparative tests of a new type of crown pick and conical pick were performed
for the constant side deflection angle value $\rho = 6°$, but in the case of standard
tangential-rotary pick, the value of side deflection angle was for the few tests much
bigger—15°, 21° and 27° respectively. The view of the measuring head, with conical
and crown pick fixed in the holder for the side deflection angle value $\rho = 6°$, ready
for the tests, are shown in Fig. 4.3.

The value of cutting depth was $g = 6, 9$ and 12 mm, cutting spacing $t_s = 12$ mm and
cutting speed 1.5 m/s. The concrete ring samples were used with external diameter
1200 mm, internal diameter 500 mm and height 450 mm, with uniaxial compressive
strength $\sigma_c = 65$ MPa.

The load of the picks, the number of rotations and the dimension of output graining
were measured during the tests. The measurement number of rotations allowed a
particular marker on the bottom of the pick handle. Also, the wear of the cutting
pick top was observed. The view of the ring sample mining for the cutting depth $g =$
6 mm and side deflection angle value $\rho = 6°$ using the new crown tool and standard
conical pick is shown in Fig. 4.4.

Fig. 4.3 A view of the measuring head, with conical and crown pick fixed in the holder for the side deflection angle value $\rho = 6°$, ready for the tests

The results of preliminary research have been satisfactory. Significant rotations (over 12 rpm) and minimal wear of the new crown tool, in the case of side deflection angle value $\rho = 6°$ were observed. For standard conical pick, the regular rotation of the tool (about 3–4 rpm) could only be observed at a side deflection angle value $\rho = 21°$.

The view of cutting pick top of new crown pick and standard conical pick, for the cutting depth $g = 6$ mm, after mining of side surface sample on the total cutting way length $L = 800$ m is presented in Fig. 4.5. In the case of the new crown pick, no visible sign of wear was observed. The working part of the standard tangential-rotary pick was practically destroyed. It caused a big dustiness and sparking (Fig. 4.3b).

The value of cutting tools load was presented in Table 4.1. By analysing the obtained values, we can conclude that for the depth of cut $g = 6$ and 9 mm, the average values of the forces acting on the tangential-rotary pick and crown pick are comparable. At a depth of $g = 12$ mm, the forces acting on the special crown pick are approx—20% higher. However, the variability of these forces is much smaller for crown picks in each case. The maximum values of the recorded forces for the special crown pick are 15–25% lower. It can be observed on the example courses of the values of the forces (Fig. 4.6) acting on the tested tools during the cutting of the

Fig. 4.4 A view of the ring sample mining for the cutting depth g = 6 mm and side deflection angle value $\rho = 6°$ using the new crown tool and standard conical pick

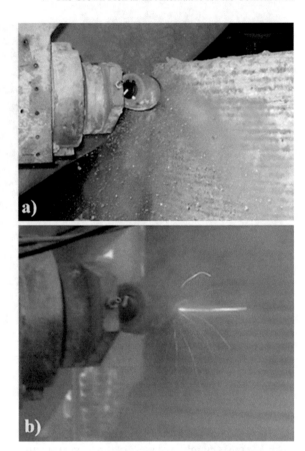

Fig. 4.5 A view of cutting pick top of new crown pick (**a**) and standard conical pick (**b**), for the cutting depth g = 6 mm, after mining of side surface sample on the total cutting way length L = 800 m

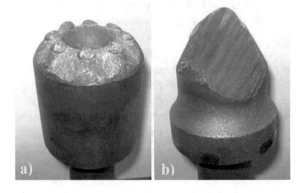

Table 4.1 The average value of the forces acting into conical and crown picks during concrete sample mining

Force type	Cutting depth g, mm					
	6		9		12	
	Conical pick	Crown pick	Conical pick	Crown pick	Conical pick	Crown pick
Pressure force Pd [kN]	7.91	6.78	10.56	11.97	15.85	19.61
Tangential force Ps [kN]	6.93	5.37	8.17	7.28	11.32	9.96
Side force Pb [kN]	1.48	1.19	1.75	2.23	2.13	3.03

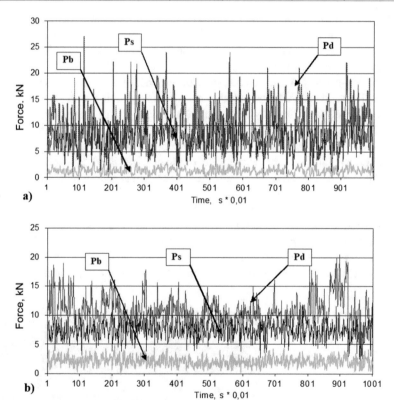

Fig. 4.6 The example courses of the values of the pressure force Pd, tangential force Ps and side force Pb acting on the tested tools during the cutting of the concrete sample at the cutting depth g = 9 mm: **a** conical pick, **b** crown pick

Table 4.2 The share of grain fractions in the excavated output obtained during concrete sample mining with conical and crown picks at a cutting depth of g = 12 mm

Grain dimensions of the obtained output (mm)	Type of cutting tool	
	Conical pick (%)	Crown pick (%)
< 5	28.6	27.4
5–20	42.7	64.3
> 20	28.7	8.3

concrete sample at the cutting depth $g = 9$ mm. The opposite trend was observed only in the case of side force.

For cuts with a cutting depth of $g = 12$ mm, measurements of the grain size distribution of the obtained output were carried out. A randomly selected amount of 1000 g was selected from the received output for each of the performed tests. The percentage of the following grain fractions for each output sample was determined by sieve analysis: grains with dimensions below 5 mm, grains with dimensions between 5 and 20 mm and grains more significant than 20 mm. The results of the performed analysis are shown in Table 4.2.

When comparing the received output visually and based on the results of the sieve analysis, it can be seen that the output obtained during cutting with a special crown pick is relatively fine but very homogeneous (almost 2/3 of the material with the same granulation in the grain range from 5 to 20 mm), compared to the conical pick. The output obtained by mining using the crown pick is shown in Fig. 4.7. In the case of standard tangential-rotary pick, the wear influence into the dimension of the grain was significant. Even a slight blunting of the pick edge caused about a 15% reduction of the fraction above 20 mm (Fig. 4.8).

During mining, with the crown pick fixed in the holder with a more excellent side deflection angle value $\rho = 12$ and $15°$, a radical increase in the number of revolutions was observed. For the side deflection angle value $\rho = 15°$, compared to cutting with the same parameters, but for the side deflection angle value $\rho = 6°$, the increase in the number of revolutions was several times. The rotation of the cutting tool also

Fig. 4.7 A view of the output obtained by mining using the crown pick at cutting depth g = 12 mm

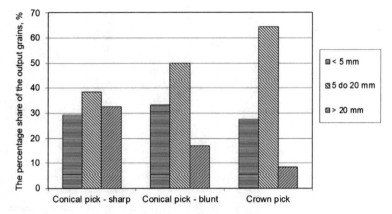

Fig. 4.8 The share of grain fractions in the excavated output obtained during concrete sample mining at a cutting depth of g = 12 mm for a sharp and blunt conical pick and a special crown pick

influenced the texture of the excavated side surface of the concrete sample. It was possible to notice the crossing of numerous cutting lines left by the cutting edges of the special crown pick, definitely more pronounced for the higher side deflection angle ro. In the case of cutting with a standard tangential-rotary pick on the sample surface, even at the side deflection angle $\rho = 21°$, the cutting lines were parallel. The view of the sample surface after cutting at the cutting depth g = 9 mm with a conical and crown pick is shown in Fig. 4.9.

4.3 Laboratory Tests of Crown Pick Solutions in Order to Select the Most Advantageous Version

In the next part of trials, several solutions of crown picks, differing in diameter and number of carbide posts as well as the pitch diameter of their location, were made for further research to determine the most favourable parameters of this tool as a function of the energy consumption, pick top-wear and the number of revolutions [1, 4].

The handle part of each pick, 75 mm long, was still 30 mm in diameter. The height of the body of the crown pick was assumed to be equal to the standard heights of the above-mentioned conical picks, i.e. 145 mm. The working part of the new version of the unique crown picks, unlike the old version, had a larger diameter at the point of contact with the surface of the holder. Because the external diameter of the contact surface of standard conical pick holders does not exceed 70 mm, this value was adopted. The diameter of the working part gradually decreased until it reached an outer diameter at the crown of 54 mm. This value corresponds to the old version of the special crown pick. The scheme of the new version of the special crown pick is shown in Fig. 4.10.

Fig. 4.9 A view of the
sample surface after cutting
at the cutting depth g =
9 mm with: **a** conical pick
with the side deflection angle
$\rho = 21°$, **b** crown pick with
the side deflection angle $\rho =$
6°, **c** crown pick with the
side deflection angle $\rho = 15°$

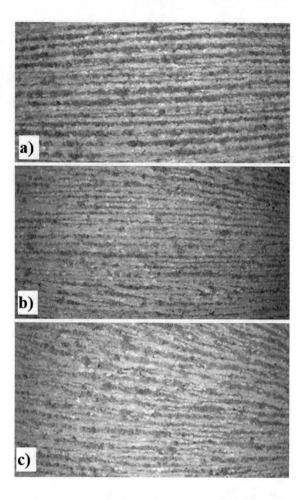

Twelve new crown picks were made—6 picks each with conical carbide inserts with 8 and 10 mm diameter. For each diameter, 3 pieces were made with the carbide insert top extended by 3 and 6 mm above the surface of the working part of the pick. A pick with the number of carbides 7 and 8 fixed on a pitch diameter of 40 and 34 mm was made for each version. There were also made 2 pieces of the picks with the ballistic carbide inserts at the diameter of 8 mm, with the number of carbides 7 and 8 on a pitch diameter of 40 mm, with the same carbide insert top extending 6 mm above the surface of the working part of the pick. The list of parameters of new versions of crown picks is presented in Table 4.3. The view of a part set of special crown picks made for the tests is shown in Fig. 4.11. Figure 4.12 presents the comparing of the old and new versions of crown pick and standard conical pick.

The comparative tests of the new versions of crown pick were performed on the same laboratory test stand for the constant side deflection angle value $\rho = 9°$, cutting depth $g = 10$ mm, cutting spacing $t_s = 12$ mm and cutting speed 1,5 m/s.

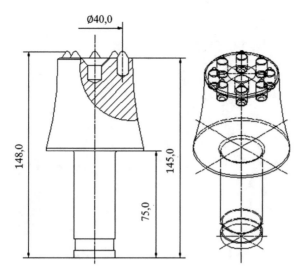

Fig. 4.10 The scheme of the new version of the special crown pick

Table 4.3 The list of parameters of new versions of crown picks

Pick number	Type of carbide insert	Diameter of carbide insert (mm)	Number of carbide insert	Pitch diameter (mm)	Carbide edge extension (mm)
1	Conical	Ø8	7	Ø40	3
2	Conical	Ø8	8	Ø40	3
3	Conical	Ø10	7	Ø40	3
4	Conical	Ø10	8	Ø40	3
5	Conical	Ø8	7	Ø40	6
6	Conical	Ø8	8	Ø40	6
7	Conical	Ø10	7	Ø40	6
8	Conical	Ø10	8	Ø40	6
9	Conical	Ø8	7	Ø34	3
10	Conical	Ø10	7	Ø34	3
11	Conical	Ø8	7	Ø34	6
12	Conical	Ø8	8	Ø34	6
13	Balistic	Ø8	7	Ø40	6
14	Balistic	Ø8	8	Ø40	6

The same concrete ring samples with uniaxial compressive strength $\sigma_c = 65$ MPa were used. For each version of crown pick, the concrete sample mining tests were carried out until the total cutting way length was at least 800 m. The load of the

Fig. 4.11 A view of a part set of new crown picks ready for the tests

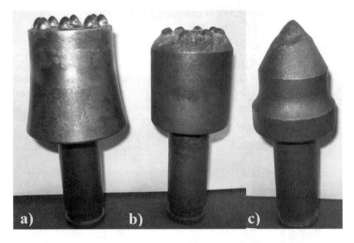

Fig. 4.12 A view of: **a** new version of crown pick, **b** old version of crown pick, **c** standard conical pick

picks and the number of rotations were measured using the same measurement methodology as before. The wear of the crown pick top was also observed.

For crown pick, no. 2 (Table 4.3), clear traces of abrasion on the upper surface of the pick working part between the carbide posts were noticed (Fig. 4.13). Therefore, special milling of the upper surface of the pick working part between the carbide inserts was made for this and other picks. The view of crown pick no. 4 (Table 4.3) with seven and crown pick no. 8 with eight carbide posts, after these cut-outs milling, is shown in Fig. 4.14.

Fig. 4.13 A view of the clear traces of abrasion on the upper surface of the pick working part between the carbide posts

Fig. 4.14 A view of crown pick no. 4 and crown pick no. 8 with special cut-outs milling

The geometry parameters had a significant influence on the load of the crown tool. The most favourable results were obtained with the use of a crown pick with eight carbide conical posts with a diameter of 8 mm, fixed on a pitch diameter of 34 mm, with the carbide insert top extended by at least 6 mm above the surface of the working part of the pick.

Increasing the pitch diameter of the carbide inserts arrangement from 34 to 40 mm caused a significant increase in the value of the forces, depending on its type—pressure, tangential or side force, from a dozen to several dozen percent. In comparison, increasing the carbide inserts diameter from 8 to 10 mm or changing the number of the carbide posts from 8 to 7 had a negligible effect on the value of the load on the tool. However, it influenced the number of tool revolutions and their wear. The larger diameter of the carbide posts resulted in a decrease in the number of pick revolutions by about 20–25% and noticeable traces of carbide wear. The lower number

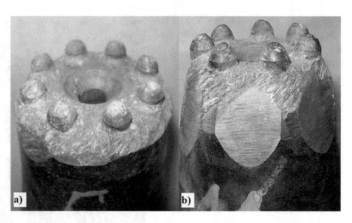

Fig. 4.15 A view of the crown pick wear after sample mining on the total distance of 1200 m: **a** one of the worse versions of crown pick (no. 5), **b** the best version of crown pick (no. 11)

of carbides, with comparable visible wear of pick top, generated a decrease in the number of pick revolutions by about 10–15%. Figure 4.15 presents the view of the best version of crown pick (no. 11 Table 4.3) and one of the worse versions (no. 5 Table 4.3) after the laboratory stand tests.

As a consequence of the research, the best solutions for the new crown pick for the following field tests have been chosen. The best results (lowest tool load, low wear and largest number of tool rotations) were obtained for the crown pick with 8 carbide inserts with a diameter of 8 mm, fixed on the pitch diameter equal to 34 mm. Slightly worse results were obtained for the crown pick with 7 carbide inserts with a diameter of 8 mm, fixed on the same pitch diameter. These two pick solutions were used in the next part of the research.

4.4 Field Tests of the New Solution of the Crown Pick

The positive results of the stand tests prompted the carrying out of mining tests with new crown picks mounted on the mining heads of roadheaders. It was planned to perform, in the first sequence, field tests in one of the limestone opencast mines with the use of a light type roadheader with a transverse mining head driven by a 132 kW engine.

The cutting head of this roadheader did not require major structural modifications related to the assembly of selected solutions of new cutting tools. However, a geometrical and kinematic analysis was carried out to check the influence of mounting crown picks on the correctness of the roadheaders operation [1, 4].

For the conical picks shown in Fig. 4.1a, used for mining limestone in the selected opencast mine, the diameter measured at the tip of the pick mounted in a holder welded on the head body with the largest radius of 240 mm is approximately 746 mm. For these parameters, the possibility of replacing the conical pick with the crown pick directly in the holders was carried out. It was proposed to use the most advantageous version of a special crown pick, with the diameter of the arrangement of the carbide posts 34 mm, the upper diameter of the working part 55 mm, the diameter of the conical type of carbide inserts 8 mm, the carbide inserts top extending 6 mm above the surface of the working part of the pick and pick height 148 mm.

By mounting a special crown pick with the parameters above in the same tool holder as for the standard conical pick, the obtained diameter value was 758 mm measured on the most extended carbide tip. This case is schematically shown in the diagram in Fig. 4.16. The difference was 12 mm.

Depending on the tool layout, this may result in a slight torque value increase of 2.5–5% from the cutting resistance for the individual tools. A greater risk is

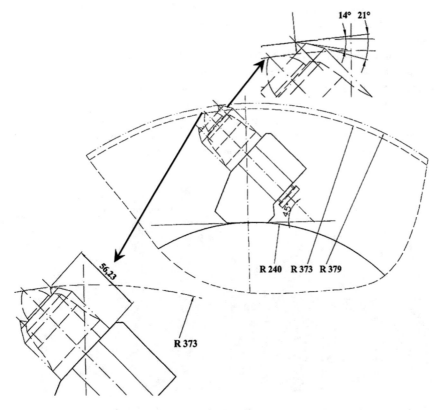

Fig. 4.16 Diagram for determining the value of the diameter measured at the tip of the conical pick and a special crown pick, the height of the working part of the crown pick and the value of the clearance angle for the carbide post edge and the skewing plane of a special crown pick

the possibility of increasing the cutting depth of the crown pick. According to the previously obtained results, the load of the crown pick in the case of mining high-strength rocks maybe even 20% higher at the cutting depth $g = 10$ mm than in the case of conical picks. This case may result in a greater demand for power or reduce the cutting speed.

Based on the diagram in Fig. 4.16, it was determined that to obtain the values of the diameters measured on the most extended carbide tip for the crown pick, similar to the values obtained for the conical picks, the working part of the crown pick should be about 56 mm high, comparing to the current value of 70 mm. This problem means the necessity to develop a new version of the special crown pick with the working part of the crown pick of 50 mm height and a total pick height not exceeding 131 mm [1, 3].

The clearance angle value for the applied sintered carbide posts and the angle for the inclined part of the working part of the crown pick were also determined. According to the scheme in Fig. 4.16, the clearance angle α of the tip of the carbide post is approximately 14°, while the angle between the inclined part of the crown pick working part and the mined rock surface is even more significant and amounts to 21°. The condition for cutting is that the clearance angle $\alpha > 0$. This condition is met.

In addition, it was checked whether the arrangement of such crown tools would not cause a collision in the present picks arrangement on the roadheaders cutting head body. The analysis was performed for the arrangement of holders with the crown pick as on a standard organ. Also, constraints were established so that the contour of the pick edges tops coincided with the contour of the standard conical tools. The 3D model view of the holder with the new solution of the crown pick and the 3D cutting head model with the set of crown picks are shown in Fig. 4.17.

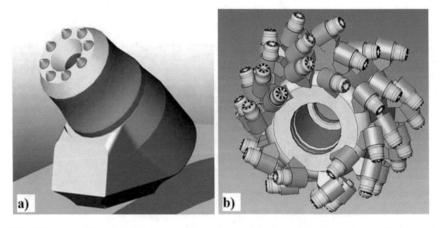

Fig. 4.17 A spatial model of the new solution of a special crown pick in a standard holder (**a**) and a spatial model of cutting head with crown picks (**b**)

Fig. 4.18 A view of the new shortened crown picks, 7 and 8-carbide inserts version [1]

The analysis results made it possible to state that there were no collisions of the elements of the picks and no collisions of the picks with the holders and the body of the cutting head. The generated coverage with new crown picks is more significant than in the case of standard conical picks, while the cutting depth value for a single tool is much lower. Two sets of the new shortened crown picks, 7 and 8-carbide inserts version were made. Both pick versions are shown in Fig. 4.18.

The field tests in the limestone opencast mine were conducted as comparative mining with a cutting head equipped with conical picks and crown picks in the 8-carbide inserts version. The view of the cutting heads equipped with conical picks and crown picks is shown in Fig. 4.19.The limestone face was mined in two 8-h shifts with each cutting head. The effectiveness, efficiency of mining, grain size distribution and tool wear were compared. The view of the limestone face mined with the cutting

Fig. 4.19 A view of the cutting heads equipped with: **a** conical picks, **b** crown picks

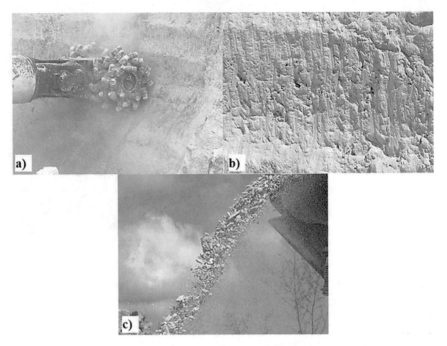

Fig. 4.20 A view of the cutting head of the light type roadheader, equipped with special crown picks, **a** while mining the limestone face, **b** the obtained surface of the limestone face, **c** the output graining

head with crown picks, its surface and the obtained excavated output is shown in Fig. 4.20 [1, 3].

In the case of mining with the cutting head with crown picks, the mining efficiency was from 34 to 44 m³/h. This value was comparable to that obtained in the case of the cutting head with conical picks.

During mining, no large dustiness was observed around the cutting head. The share of grain fractions in the output ranged 0–5 mm, 5–15 mm and over 15 mm was carried out. For the cutting head with standard conical picks, the grain fraction share below 5 mm was 27.6%, the grain fraction shares 5 to 15 mm—30.1%, and the grain fraction above 15 mm—42.3%. In the case of the cutting head with crown picks, the results differed quite significantly. The obtained share of the grain fraction below 5 mm was 21.3%, the share of the grain fraction 5 to 15 mm—44.1% and the grain fraction above 15 mm—34.6%. The output obtained during mining was homogeneous (Fig. 4.20c).

After the mining, on the surface of the limestone face, a grid of cuts left by the edges of cutting picks was visible. These cuts looked different than in the case of a cutting head with conical picks. A much shorter distance separated them from 1.5 to 4 cm, and the depth ranged from 8 to 18 mm. The view of a face surface after mining with the cutting head with crown picks is shown in Fig. 4.20b.

Fig. 4.21 A view of the special crown picks after the limestone face mining during the two working shifts [1]

After the mining was completed, no signs of wear of the carbide inserts of the crown picks and their bodies were noticed. The only adverse effect was the hole contamination in the central part of the working part of the pick with fines and dust. Most of the conical picks showed signs of progressive wear at comparable operating times, mainly non-symmetrical grinding of the pick edge. Figure 4.21 shows the view of the special crown pick on the cutting head after limestone excavation during two working shifts.

Based on the interview with the roadheaders operator, information was obtained that compared to working with a cutting head equipped with the conical picks, the roadheader with a cutting head with special crown picks was more challenging to operate. It was possible to observe its significant vibrations and repeated stopping of the mining head due to the high value of its speed of movement. It was necessary to operate the mining head very gently and move it at low speeds to prevent this from occurring.

4.5 Industrial Tests of the New Solution of the Crown Pick

Like the last one, new tools were tested under underground conditions in one of the hard coal mines. The new crown picks were to be mounted on the mining head of a medium roadheaders with the power of the driving motor of the 200 kW cutting head. After conducting a similar analysis as before, it was found that the 7-carbide inserts version of crown pick can be mounted in the holders of a standard cutting head of the used roadheaders [1, 4].

The roadheaders mined a sandstone gallery face with a compression strength from 100 to 120 MPa. The disadvantage of the excavated rock was its very high homogeneity, almost complete lack of layering and a considerable proportion of

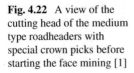

Fig. 4.22 A view of the cutting head of the medium type roadheaders with special crown picks before starting the face mining [1]

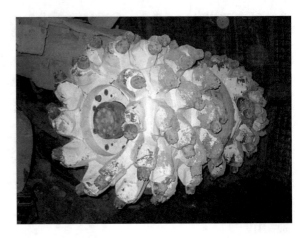

quartz grains in the sandstone. It was the reason for minimal daily advance and high wear of cutting tools, up to 60 pieces per shift.

The view of the roadheaders cutting head with new crown picks ready to work is shown in Fig. 4.22. The test was performed using the same operator who operated the roadheader while cutting the face with the cutting head with conical picks. During the first mining attempts, a very high instability of the roadheader was noticed, as well as the vibrations of the machine and the temporary stopping of the engine's rotation driving the mining head.

The roadheaders operator monitored the active power consumption at a level exceeding 200 kW. The reason for unstable work and high-power consumption was a too large depth of cut by single tools, caused by the too high speed of movement of the mining head. The system for controlling the movements of the cutting head did not ensure the minimum value speed of the head moving, which would allow for the cutting depth below 20 mm [1].

After the tests, the mining head was visually inspected. A huge number of damaged picks were observed. It was not damage caused by abrasion or grinding. These were damage caused by the impact of the carbide posts and their chipping or breakout. A view of a few crown picks' with chipped and broken carbide inserts is shown in Fig. 4.23.

Based on the obtained test results, it can be stated that the use of crown picks is pointless for such a mining technology with a roadheader. Guiding the mining head during the tests with too much moving speed caused excessive penetration of the tools into the rock massive and excessive loading of the mining head motor and generation of vibrations. First, it is necessary to apply a system that allows for a gentle and slow movement of the cutting head to ensure appropriate depths of the cut. This way will allow for the stable operation of the machine without dynamic changes having a very detrimental effect on the work efficiency and service life of both the cutting tools and the individual components of the machine.

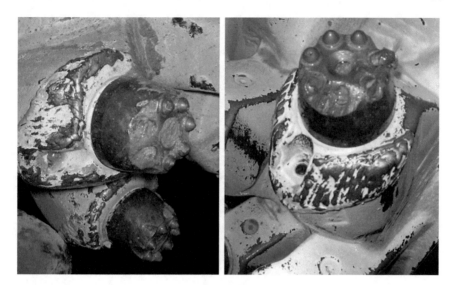

Fig. 4.23 A view of a few of the crown picks with chipped and broken carbide inserts

References

1. Kotwica, K., et al.: Collective work—Selected problems of mining, transport and dressing of highly cohesive rocks, Part 1, p. 147. AGH UST Publisher (2016). (in Polish)
2. Kotwica, K., Klich, A., Kalukiewicz, A., Gospodarczyk, P.: Miningtool. Patent of the Polish Patent Office No. PL 182135, Warszawa (2001)
3. Kotwica, K., Klich, A.: Machines and Equipment for Roadways and Tunnels Mining, p. 313. Instytut Techniki Górniczej KOMAG, Gliwice (2011). (in Polish)
4. Kotwica, K.: Hard rock mining—cutting or disk tools. IOP Conf. Ser.: Mater. Sci. Eng. 545(1), 1–11 (2019). Article no. 012019

Chapter 5
Lubricated Holder for Conical Pick

Abstract This chapter describes the new solution of a lubricated holder as an alternative for the standard holder of conical pick. The analysis of the results of the own tests proved that the rotation of conical picks is affected not only by the value of side deflection angle ρ, and cutting angles depending on the properties and lithology of the mined rocks and the manner of their bearing in the holders. The first part described the types and causes of conical pick edge damages occurring during hard rock mining. The internal and external spraying using high-pressure water jets to eliminate or reduce hazards occurring during rock mining with conical picks were presented. The second part presents the constructional solution of a lubricated holder for the conical picks. Based on the obtained results of the tests performed on the unique laboratory stand, the usefulness of the new lubricated holder solution has been confirmed. Comparing the number of rotations in the holder and the size and type of the pick wear in the case of the conical pick mounted in the standard holder, holder with high-pressure water jet assistance and lubricated holder, the most favourable results were obtained for the picks assembled in the lubricated holder. The holder was supplied under pressure p with 1.5% emulsion or pure water. The next part describes the laboratory trials to choose the most convenient solution for the lubricated holder. Four holders were checked: open sleeves with straight and helical grooves and closed sleeves smooth or with helical grooves. The tests were conducted on the new, unique laboratory stand, which allowed performing tests of resistance of the rotation of the tool in the holders. The holders with open sleeve with straight grooves and closed sleeve smooth were chosen. The comparing tests using the light type of roadheaders equipped with standard mining head and lubricated mining had were performed in the field and underground conditions. The results of both trials have shown the usefulness of the new lubricated holder solution for conical picks, with pure water as the lubricant. Application of water-lubricated holders increased almost twice the working time of picks and reduced the dustiness.

In Chap. 2 was described that the tangential-rotary (conical) picks require regular rotation in the holders to ensure even top pick wear. Figure 2.9 presents the very regular and symmetrical wear of the pick top, obtained with regular and numerous tool revolutions in the holder. This type of conical pick top wear is usually possible

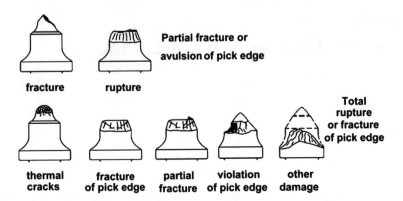

Fig. 5.1 Types of conical pick edges damages according to [5, 7]

in laboratory conditions. In practice, only the pick edges of about 1/3 of the tools wear similarly. Exemplary types of conical pick edge damages occurring during hard rock mining are presented in Fig. 5.1 [5, 7].

The decay of rotation leads to increased and very uneven wear of the working part of the pick. This type of wear is shown in Fig. 2.11. Such a tool generates numerous hazards in the form of sparking, dusting and increases the demand of power needed for mining, with a much lower mining efficiency and smaller granulation of the excavated material. The possibilities of decreasing or even eliminating these hazards are briefly presented in Chap. 1.

In Chap. 4, the new tool alternative—crown pick, for the conical picks was described to obtain much higher torque values acting on the working part of the cutting pick. The second way for increasing tool rotation in the holder is to reduce the value of the friction coefficient between the pick shaft and the inner surface of the holder.

This chapter describes the possibility of reducing the value of the friction coefficient in the holder for standard conical pick, which is shown in Chap. 4 in Fig. 4.1a.

5.1 Possibilities of Elimination or Reduction of Hazards Occurring During Rock Mining with Conical Picks

Reduction of hazards occurring during the mining of rocks with conical picks, such as sparks or dust, and increased wear of pick edges in the case of mechanical mining of rocks with milling roadheaders, is possible through the use of high-pressure water spraying. Two types of sprinkling are possible: external, the so-called water or air–water curtains, and internal, the so-called sector or individual [5, 7].

In the case of external spraying, which is simpler and cheaper to operate, the influence of the water jets is on the whole cutting head or the mining head. However, this sprinkling method does not prevent the occurrence of hazards such as sparks or dust but only limits their propagation beyond the water or air–water curtain. It also slightly reduces the wear of the cutting tool edges.

Currently, external spraying is used occasionally due to high water consumption, lower efficiency in fighting dustiness, and higher pick edge wear. Much better is the internal system applied in most of the manufactured roadheaders. It enables directing the high-pressure water jet precisely in the area of the cutting tool operation. It allows the reduction or elimination of the causes, not the effects, of threats.

According to the described in Chap. 2 pattern of rock cutting, cracks and micro-cracks get created in front of the cutting tool edge. Whereas, from the side of the application plane, the most significant factor in the wearing process of the tool edge is the friction of the bottom surface of the mined rock.

When the high-pressure water jet leads into the rock, it hydrates the rock massive. For rocks equal to or higher than the liquid limit, cohesion is very low. Also, little moistening of rock causes the limitation of cohesion forces, as well as for a rock in a wet state; the inner friction gets lowered.

Additionally, when mining a wet rock, there is a decrease in the coefficient of friction between mined rock chips and the tool. Whereas, from the side of the tool application plane, lowering the friction coefficient between the rock and the tool decreases the friction force T significantly and the amount of produced heat and the tool wear. The application of individual high-pressure water jet assistance of the mining process provides many benefits connected with the limitation of mining resistance values, increased output volume, or decreased tool wear [4, 5].

At mining with the conical pick, there are two possibilities of assistance of the tool operation with a high-pressure water jet. Figure 5.2 presents the high-pressure nozzle location—in the front of or in the rear of the cutting tool. For both locations, the nozzles are mounted on the tool axis.

For the nozzle location in front of the cutting tool, the jet distance from the tool edge should amount to 1–3 mm. Such a configuration allows the slight reduction of forces affecting the tool. The high-pressure water jet reaches the area of the cracks

Fig. 5.2 A diagram of high-pressure nozzles location in the mining with a conical pick [1, 3, 5]

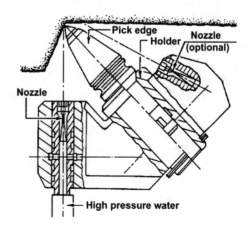

Fig. 5.3 Scheme spraying
of the cutting pick spraying
behind the pick: 1—water
supply, 2—high-pressure
water jet, 3—the highest
temperature zone, 4—the
zone with the most
significant dust formation,
5—crumpling zone,
6—crushing zone [1, 3, 5]

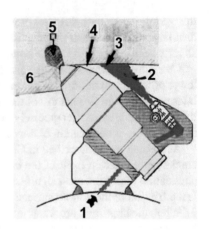

zone induced by the tool and weakens the rock structure. But the main hazardous—sparking and dustiness are generated behind the tool when the water jet does not arrive.

When the nozzle is mounted in the holder behind the tool edge, it enables sprinkling the rock in the place most endangered to spark occurrence, which may initiate ignition and the area of dust production. The sparks are practically eliminated when the jet axis hits the cut bottom not longer than 30 mm from the tool edge. This application also significantly reduces dustiness, sometimes up to 90% and more. It is shown in the diagram in Fig. 5.3.

But even this solution of high-pressure spraying does not allow protection of the conical pick rotation decay in the holder.

After even a short break in the mining process, it was repeatedly found that the conical picks had been blocked in the holders. It was caused by wet dust and small particles of mined rock which got between the pick shank and the inner surface of the holder. When wet dust dries, it completely stops the rotation of the pick.

It was the reason for starting in the AGH UST Cracow with the researches for limitation or prevention of this phenomenon. The developed solutions and results of performed tests are described in the next part of this chapter.

5.2   The Construction and Principle of Work of the New Solution of the Lubricated Holder of Conical Pick

Cutting tools, mainly conical picks, operate under challenging conditions, especially hard rock mining. The small pieces of the mined rock, dust or water used for sprinkling contaminate the zone of "bearing" of tool holders and make it difficult for conical picks to rotate in the holders. The problems of rotation depend on the kind of sliding friction.

Dry friction usually occurs in the tool bearing constructions, characterised by a relative displacement of collaborating elements with no participation of a lubricating medium and cutting down uneven surfaces' tops. Dry and semi-dry friction is based, according to general principles, on the action of adhesive forces and on overcoming elastic and deformation resistance of micro irregularities straining, i.e. on shearing metallic "joints" at the points of the accurate contact. A very high coefficient of friction is typical for "clean" metallic surfaces, which requires exceeding the yield point at the points of accurate contact [4, 5].

In this type of friction pair, the friction force is generally composed of a sum of two forces operating within the friction zone:

$$T = T_s + T_b \qquad (5.1)$$

where:

$T_s = S \cdot R_t$—the force essential to shear the irregularities of a weaker material [N],

S—the total shearing area of a given zone, [mm^2],

R_t—the experimental value of the steel shear strength, the range of value was assumed $R_t = 60 \div 140$ [MPa],

$T_b = S_b \cdot \rho_w$—the force needed to press a furrow or shear the material of a foreign matter [N],

S_b—a cross-section of a furrow or particles of a foreign matter [mm^2],

ρ_w—elementary resistance of pressing the furrow or the resistance of inclusion shearing. The value is comparable with the yield point of a softer material [N/mm^2].

Elementary extrusion resistance is proportional to the yield point of a softer material and the deformation speed.

If the link between sliding surfaces results from the fact that the irregularities and foreign matters get caught one to another or adhesive forces are active, the destruction of collaborated surfaces will take place by abrasion. Abrasive wear may be generally defined as a destruction of a large surface performed by a material harder than the surface itself.

Dry, semi-dry, and mixed friction could have occurred in the conical picks holders used on the cutting heads. Mixed friction will occur when the tangential tools mounted in the mining head are sprinkled with water or emulsion. Such friction reduces the process of abrasive wear and makes it easier for the elements to relocate one towards another.

As has been already mentioned in Chap. 2, the bearing of tangential rotary tools in the mining heads takes place either directly in the holders or the holders with hardened exchangeable sleeves. The tools are set up in the holders at the given angles to maintain their proper operation. Such a mounting way of the tool assembly in the holder generates dry friction that deteriorates the tool rotation and causes their premature asymmetric wearing.

In AGH UST in Cracow, some attempts aimed at bearing the tangential rotary tools in the holders towards changing dry or semi-dry friction for mixed or semi-fluid friction to make the tools rotate. Due to the limited space associated with the construction of the tools and holders and low rotational speed, fluid friction cannot be acquired in any case [4, 5].

In these types of bearing, the requirements for Sommerfeld number So (Eq. 5.2) should be met. Considering part of the elements of this equation—tool angular velocity ω and surface pressure in the bearing sleeve p, even when very viscous lubricating media are applied, these requirements will be not met.

$$So = \frac{\eta \cdot \omega}{p \cdot \Psi^2} \tag{5.2}$$

where:

η—viscosity of a lubricating medium [Pa·s],
ω—tool angular velocity [1/s],
p—surface pressure in the bearing sleeve [MPa],
Ψ—relative clearance for slide bearings on fluid friction should range from $\Psi = 0.0006$ to 0.015 as it depends on the bearing load and pin rotational speed.

Some efforts to apply exchangeable slide bearing sleeves in the holders were applied. The sleeves of hardened materials, self-lubricants, plastics based on Teflon or materials reinforced with carbon fibres and composites were analysed.

Due to the difficulty in sealing of bearing joint operating in callous conditions, standard abrasive wear index was taken in the analysis into account. Compared with bronze, the alternative material for the sleeves can be polymers with filling material or reinforced with fibres.

The bearings of rotary-tangential tools tend to operate under unfavourable conditions, where the sliding speed is insufficient and the surface pressure—high. It is then challenging to provide proper lubrication, and additionally, the output and abrasion products press themselves between frictional surfaces.

The advantage of those materials is that they are not susceptible to corrosion and may be lubricated with water, water with impurities or wear products, which push into a soft material of a bearing sleeve.

Going further in AGH, UST Cracow has attempted to find the methods to replace sliding friction with rolling friction at the bearing of tool handles in the holders. The way of rolling bearing is determined, in a limited constructional space, by a rest load-carrying ability of the rolling bearings or their combinations possible to be mounted. Rest load-carrying ability results from Hertz contact pressures and, as a matter of fact, from the admissible values of contact pressures k_{dh}, which depend on the state of the material effort and hardness of the elements in the contact. The values of contact pressures admissible for rolling associations account for $k_{dh} = 2600$ MPa, and globular associations, $k_{dh} = 3500$ MPa, at the element hardness in the contact $FH_{HB} \geq 60$ HRC ≥ 700 HB.

Fig. 5.4 The constructional solution of the holder for rotary-tangential picks with rolling bearing: 1—tool holder, 2—pick handle, 3—globule, 4 - needle bearing, 5—sealing [4, 5]

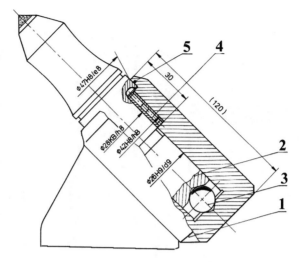

Based on the results of the strength analysis of the tools and holders most often used in the mining industry, the alternative method of rolling bearing for rotary-tangential tools was prepared. The sample constructional solution is presented in Fig. 5.4 [4, 5].

This solution applies needle bearings as the elements transmitting the load transverse to the tool axis and the globule for transferring the axial load. At the rolling bearing of rotary-tangential tools, it is not required to select bearings regarding mobile load-carrying ability because they rotate slowly, their life is limited, so the relationships resulting from the Hertz contact pressures, i.e. $\sigma_H < k_{dh}$ seem to be sufficient. Therefore rolling elements may be much more loaded, and thus the rolling joints—miniaturized.

But the disadvantages of this solution are complicated construction, the higher cost of manufacturing and replacement of the bearing requiring special conditions that allow for the correct assembly of the bearing.

Based on the information mentioned above and on the analysis of constructional possibilities and own preliminary investigation, the alternative solution of conical pick holder, using the methodology of multivariant consideration, was prepared at the AGH UST in Cracow. Dry friction between the tool handle and the inner surface of the holder was replaced by semi-dry or mixed friction. These kinds of friction develop due to a partial separation of irregularities' tops, their moistening or covering with the emulsion, and application of self-lubricants and plastics with some additives on the bearing sleeve make it easier to lubricate. The constructional solution of the new holder for the standard conical pick is shown in Fig. 5.5. The semi-dry or mixed friction was obtained by forced lubrication of the exchangeable sleeve in the holder [4–7].

In the proposed solution, a lubricating agent (low-percentage oil–water emulsion or pure water) is distributed under pressure through the hole of ferrule 2 to four grooves made in sleeve 3, mounted in the tool holder 1. The new solution of the

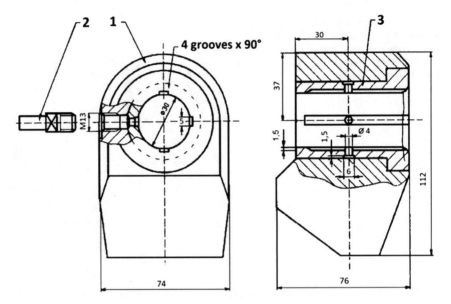

Fig. 5.5 Scheme of the lubricated holder for the standard conical picks: 1—tool holder, 2—delivery ferrule, 3—sleeve [5, 7]

lubricated holder has been submitted to the Polish Patent Office and is covered by patent protection No. PL 209806 [2].

To verify the effectiveness of the developed solutions, they were made and subjected to comparative tests on a laboratory test stand, previously presented in Chap. 4. The course and results of these tests are described below.

5.3   The Comparative Laboratory Tests of the Standard and Lubricated Conical Pick Holder

To assess the effectiveness of new constructional solutions of holders of tangential rotary tools, one copy of both holders has been made with forced lubrication and with the rolling bearing of the tools. Such the solutions were examined on a unique laboratory stand designed to test individual mining tools, described in Chap. 4 and shown in Fig. 4.2.

As was mentioned, the laboratory stand allows mining an artificial concrete sample along its side surface at constant parameters such as cutting depth g, cutting spacing t_s and the cutting speed v_s and at the given tool setting angles [5, 7].

The pick holder was welded to the special plate. The view of the plate with the attached lubricated holder is shown in Fig. 5.6. Through movable mounting with the measuring head, this plate can be turned in relation to the axis of the measuring head. The cutting pick is mounted in the holder at a setting angle $\delta = 45°$. The side

Fig. 5.6 View of the plate
with attached lubricated
holder ready to assembly on
the measuring head

deflection angle value ρ can change in 3° increments, in the range of $\rho = 0°$ up to
45°.

The tests were aimed to compare new solutions of the holders—lubricated and
with rolling bearing, with standard holder for conical picks with the exchangeable
sleeve. All trials were carried out using conical picks shown in Fig. 4.1a with carbide
inserts of 22 mm diameter. The number of tool rotations and the size and character
of their wear was measured. During the trials, the cutting depth was changed from g
= 9 to 12 mm, and the angle of side deflection ρ varied within the 0–45° range. The
resistance of a rock sample to uniaxial compression $\sigma_c = 65$ MPa, cutting speed v_s
ca. 1.25 m/s, and the cutting spacing $t_s = 12$ mm were kept constant.

The sample was mined along with its total height. The special marking gauges
were positioned on the bottom of the tool handles—these marks allowed for
measuring revolutions by observation and counting their number per unit of time. The
time interval was 1 min. This method of measurement ensured its accuracy within
¼ rpm. In addition, this part of the pick handle during the tests was filmed with a
video camera, which made it possible to control the measurement after the tests were
carried out. The movie could be played at 25 frames per second. The measurement
of the number of revolutions was started with the fixed work parameters of the tool,
i.e. when the cutting was carried out at a constant depth and spacing of cutting.

The lubricating holder was supplied either with 1.5% water–oil emulsion (first
case) or pure water (second case), and the value of supply pressure was constant $p =$
1.5 MPa for both cases. The view of the mining process with the picks mounted in a
standard holder and new lubricated holder during the trials is presented in Figs. 5.7
and 5.8.

In the first stage of the research, tests were carried out for the standard conical
pick, assembled sequentially in the standard, lubricated and rolling bearing holders.
For each case of trial, the new pick was used. The influence of the side deflection
angle value on the number of revolutions of the tool was examined. The angle of the
side deflection angle of the pick ρ assumed the following values: 0°, 6°, 12°, 18°,
36° and 45° in the case of the standard holder and only 0°, 6° and 12° in the case of

Fig. 5.7 View of the conical
pick during the mining of a
concrete sample, placed in
the standard holder at the
side deflection angle $\rho = 36°$

Fig. 5.8 View of the conical
pick during the mining of a
concrete sample, placed in
the lubricated holder at the
side deflection angle $\rho = 12°$

both others holders. The rock sample was mined with a constant cutting pitch of $t_s = 12$ mm. The tests were carried out for two different depths of cut, $g = 9$ and 12 mm. The results obtained during the tests are presented in Tables 5.1 and 5.2 and Fig. 5.9. The conical pick mounted in the standard holder was turned out that the first signs

Table 5.1 Number of
revolutions of the conical
pick, placed in the standard
holder in the function of the
cutting depth g and the side
deflection angle ρ

Side deflection angle ρ, °	Number of tool revolutions n, 1/min	
	$g = 9$ mm	$g = 12$ mm
0	0	0
6	0	0
12	0	0
18	0.50	0
36	1.25	0.50

Table 5.2 Number of revolutions of the conical pick, placed in the lubricated holder in the function of the cutting depth g, type of lubricant fluid and the side deflection angle ρ

Side deflection angle ρ, °	Type of lubricant fluid	Number of tool revolutions n, 1/min	
		g = 9 mm	g = 12 mm
0	1.5% emulsion	1.25	0.50
6		6.00	1.50
12		15.5	11.00
0	Pure water	0.75	0.25
6		4.75	1.25
12		12.75	8.5

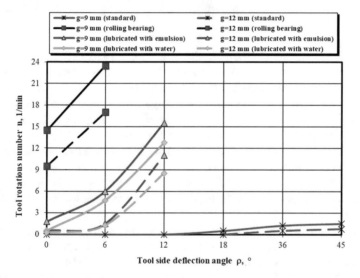

Fig. 5.9 An influence of the pick side deflection angle ρ on the number of its rotations in the function of cutting depth g and type of holder [1, 5, 7]

of the tool's rotation were observed only at the side deflection angle value $\rho = 18°$. However, even deflection of the tool by the angle $\rho = 36°$ did not bring the expected results, i.e. the tool still did not show smooth rotation (its rotations were random) and often stopped in the holder.

The additional test was made for the lubricated holder when the pick attached to this holder mined the sample without lubrication, and its rotation in the handle stopped. For the side inclination angle $\rho = 6°$, after turning on the emulsion supply of the holder, the pick began to rotate again almost immediately. Figure 5.10 shows the situation at intervals of every 0.5 s from switching on lubrication.

In the case of the holder with rolling bearing, even at zero deflection angle of the tools, they rotated in the holder. Also, satisfactory results were obtained for the

Fig. 5.10 View of conical
pick attached in the
lubricated holder during the
cutting of rock sample every
0.5 s since the beginning of
lubrication supply [3, 5, 7]

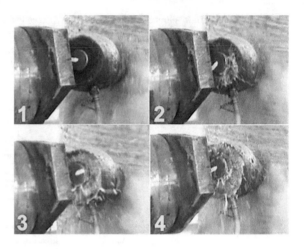

holder with lubrication. However, in this situation, the side deflection angle ρ had
to be at least 6°. The cutting depth g from 9 to 12 mm influenced the number of
rotations, mainly for the standard and lubricated holders. The number of rotations
was about 2.5 times less. The lubricant change from emulsion to pure water had
a negligible impact on revolutions. In the case of pure water, the decreasing number
of revolutions was not more significant than 15–20%.

The best effect was obtained for the holder with a rolling bearing. However, the
complicated construction of rolling bearing, requiring high precision at assembling
processes, may not be accepted for application on mining heads in underground
conditions. It was because only the lubricated holder solution was tested for the next
part of the trials.

During mining with tools mounted in holders lubricated by both water and emulsion, the more or less regular rotation of the tools was observed. It caused regular
and slight wear of the edges of the tools. In the next part of the trials, for comparison
of the pick edge wear, the tests of mining with the use of a standard holder and lubricated holder were performed. The quantitative and qualitative pick edge wear was
measured. Additionally, the test with picks attached in the standard holder assisted
with a high-pressure water jet from the front, and the rear of the tool was carried
out. The water pressure value was $p = 45$ MPa and nozzle diameter d of 0.8 mm.
The lubricated holder was supplied with 1.5% emulsion with a pressure of about $p
= 1.0$ MPa and flow rate Q of 1 dm^3/min.

The concrete ring samples at the uniaxial compressive strength $\sigma_c = 105$ MPa
were mined at the side surface at the cutting depth $g = 9$ mm, the cutting speed $v_s
= 3$ m/s, at side deflection angle $\rho = 9°$ and the total path of cutting at 2500 m. The
same as the earlier conical pick was used, the new one for each trial. A view of the
picks during mining tests without and with assistance is presented in Fig. 5.11.

The results of both quantitative and qualitative wear are significant. For the pick
attached in the standard holder without water jet assistance, after only a few dozen
seconds, the stabilization of the tool position was noticed. It did not make any, even

Fig. 5.11   A view of mining a concrete sample with a conical pick attached in the standard holder: **a** without high-pressure water assistance, **b** with high-pressure water assistance from the front, **c** with high-pressure water assistance from the rear according to [5, 7]

minimal rotational movements. After about 5 min, the pick edge was turned red hot (Fig. 5.12). After reaching a cutting path of about 800 m the total destroying of the working part of the pick was observed. For the other picks, the desired cutting path length has been reached. The view of pick edge wear is shown in Fig. 5.13, and the volumetric value of the wear presents Figs. 5.14 and 5.15.

Comparing the wear of the pick edge for the pick attached in the lubricated holder and the pick assisted with a high-pressure water jet can be concluded that the first signs of asymmetrical wear occurred in the case of water-assisted picks. Figure 5.16 presents the pick edge profile obtained using an optical microscope before and after cutting for pick assisted with the water jet from the rear and pick attached in the lubricated holder. In the case of the pick mounted in the lubricated holder the regular and low pick edge wear was obtained.

Fig. 5.12   A view of the edge of the pick attached in the standard holder, after concrete sample mining [5]

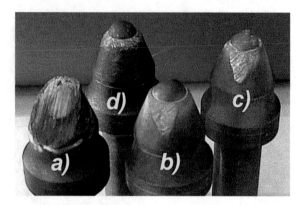

Fig. 5.13 A view of picks edges after mining tests: **a** without high-pressure water assistance, **b** with high-pressure water assistance from the front, **c** with high-pressure water assistance from the rear, **d** pick attached in the lubricated holder [3, 5, 7]

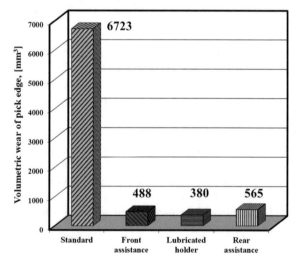

Fig. 5.14 Volumetric wear of tools edges after mining tests without and with high-pressure water assistance and in lubricated holder [3, 5, 7]

5.4   Laboratory Tests of Lubricated Holder Solutions of Conical Picks to Select the Most Advantageous Version

The tests described in the previous chapter showed that ensuring regular rotation during cutting has the most beneficial effect on the quantitative and qualitative wear of the conical pick edge. Compared to the results obtained for the mining with a high-pressure water jet assistance in front of and behind the tool edge, the value of the

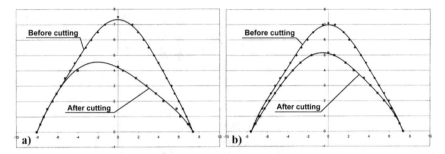

Fig. 5.15 The profiles of conical picks edges before and after cutting: **a** pick with the high-pressure water jet assistance from the front, **b** pick attached in the lubricated holder [5, 7]

Fig. 5.16 The scheme of developed solutions of lubricated holders, with attaching plates: **a** with the open sleeve with straight grooves, **b** with the open sleeve with helical grooves, **c** with the closed, smooth sleeve, **d** with the closed sleeve with helical grooves according to [5]

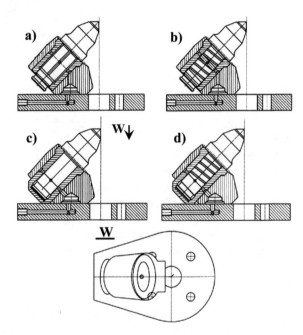

volumetric wear of the pick mounted in the lubricated holder was about 30–50% lower, and the nature of the edge wear was very regular.

Therefore, several special, new solutions of holders with sleeves, adapted to the supply of a lubricant medium and possible to be mounted on the mining head of roadheaders, were developed for further research. Also, designs of plates for the holder's assembly with a special solution of the supplying of the lubricated medium were worked out. The holders allowed the mounting of the standard conical picks used in the first part of trials. After conducting analyses, there were chosen solutions of the open sleeve with straight and helical grooves and a closed sleeve with helical grooves and without them. The lubricating medium was delivered via the channels

Fig. 5.17 A view of two
executed sleeves open sleeve
with straight and helical
grooves and a holder
mounted on an attaching
plate, ready for tests

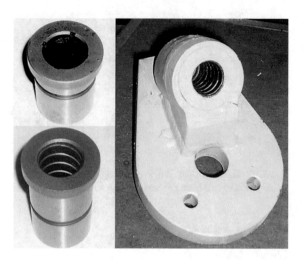

drilled into the plate and the holder body to the four holes made every 90° on the
sleeve circuit [5]. The scheme of developed solutions of lubricated holders, fixed by
welding on the plates, is shown in Fig. 5.16. A view of two executed sleeves and a
holder mounted on an attaching plate, ready for further examinations, is presented
in Fig. 5.17.

The tests performed on the laboratory stand described in Sect. 4.2 require much
workload, cost, and time. For the elaborated solutions of holders testing, it was
suggested to develop a new, unique laboratory stand which would allow, at minimal
costs, performing tests of resistance of the rotation of the tool in the holders. A scheme
for such a stand is presented in Fig. 5.18. The view of the stand shows Fig. 5.19.

This stand allows to determine the resistance to rotation of the picks in the holders,
and more specifically, to measure the value of the torque causing this rotation. Additionally, for specific measurement conditions, it was also possible to estimate the
value of the friction coefficient between the surface of the working part of the pick
and the front surface of the sleeve.

The base of the stand is frame 1, with liftable plate 2 mounted on the articulation.
The construction of plate 2 allows the assembly of the plates with holders and picks,
shown in Fig. 5.19. Cut-outs on the liftable plate enable the movement of holders
along with the plate. In the front part of the frame, a roller set 3 is mounted, with
the ability to move. Whereas the upper part of the frame is welded the guide 5 of the
weight set 6 [5, 7].

The weight set 6 moves on it, possibly rotating due to side plates 7 of the weight
set. A weight 8 with the desired mass is placed in the weight set, and it presses with
controlled force the tested tool 4 located in the holder. It can press the tool on its
axis or from a side at a certain angle. Additional stand elements are unique clamping
rings for conical picks fixed on the working part of the pick.

The trials relied on mounting on the liftable plate the examined set of the plate with
tool holder, a conical pick and the ring. There was a string wound on the ring, scrolled

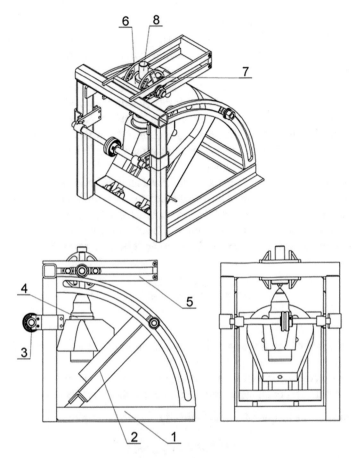

Fig. 5.18 The unique laboratory stand scheme for the research of the resistance of rotation of the picks in the holders (description in the text) according to [5, 7]

Fig. 5.19 The special laboratory stand for the research of the resistance of rotation of the picks in the holders

by the roller 3, and ended with a hook with the attached weight of the controlled mass. After loading the pick tip with a weight of a certain mass and demanding direction, the value of mass attached to the hook was increased until the tool rotation was observed in the holder.

The increase of the value of the mass of the load suspended on the string was proposed by slowly adding water to the empty plastic bottle. This solution made it possible to determine the torque value at which the movement of the pick in the holder occurred. The mass of the load was measured by weighing it on a scale with an accuracy of 1 g. At least 5 tests were performed, and in the case of significant differences in measurements—at least 7 tests.

Knowing geometrical parameters of the sleeve and holder in which the sleeve was pressed in, as well as the value of the force affecting the string causing rotation of the conical pick in the holder, it is possible to calculate the torque and friction coefficient value between the pick working part and the sleeve surface. The general formula to assess the friction coefficient value will have the following form [5]:

$$\mu = \frac{M \cdot F}{Q \cdot S_o} \tag{5.3}$$

where:

M—the value of the torque at which the rotation of the tool in the holder took place,
F—surface contact area of the working part of the pick and holder with sleeve,
Q—the value of a polar static surface moment concerning its gravity centre.

The value of the torque M was calculated using the formula:

$$M = P \cdot r \cdot \sin \alpha \tag{5.4}$$

where:

P—the force of gravity affecting the mass of the bottle with the water, suspended on the string, calculated from the formula: $P = m_{b+w} \cdot g$, N
m_{b+w}—the measured weight of the bottle with the water, kg
g—gravitational acceleration, m/s^2
r—radius of string winding on the ring, m
$\sin \alpha = 1$, because the angle between the direction of P force action and the radius r has 90°.

The contact surface area of the tool and holder was calculated using the value of the pick shank diameter and the diameter of the sleeve or working part of the pick (if it is smaller than the sleeve diameter). The value of vertical force loading the pick depends on the weight set and tool mass. Whereas the value of a polar static surface moment concerning its gravity centre S_o was calculated from the formula:

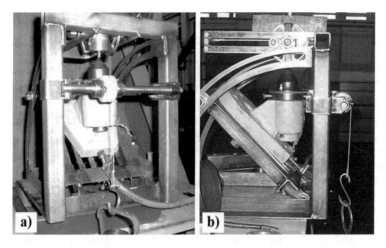

Fig. 5.20 The view of a holder during the tests with **a** the sleeve with straight grooves, **b** the closed smooth sleeve

$$S_0 = \frac{2}{3}\pi\left(r_1^3 - r_2^3\right) \tag{5.5}$$

where:

r_1—radius of the sleeve or tool head (if it is smaller than the sleeve diameter),
r_2—radius of the tool shank.

Calculating the friction coefficient value between the surface of the working part of the pick and the surface of the sleeve and holder is possible only if the load acts towards the tool axis and friction resistance between the pick shank and sleeve are omitted. In other cases, it is only possible to compare calculated values of the torque at which the rotation of the tool in the holder took place.

For each holder, there were several tests using picks with a shank of 30 mm diameter but of different diameters of the working part of the pick, without or with lubrication of the holder with low-grade emulsion or water at a pressure of about $p = 1.5$ MPa and flow rate Q from 0.5 to 1 dm^3/min. The location and value of the force loading the tool were changed as well as the condition of the surface between the tool working part and the sleeve (it was polluted with tiny stone dust).

Compared to tests performed with holders without lubrication, the best results were obtained in the case of the holder with a closed smooth sleeve. Even clean water applied at the pressure of p 1 MPa and flow rate Q of 1 dm^3/min caused a decrease of the rotation resistance from 50 to 80%. Lubrication with the emulsion at the same parameters decreased the resistance by 5 times. In the case of the holder with an open sleeve with straight grooves, water lubrication made the value of the friction coefficient fall by about 25–40%.

Table 5.3 The averaged values of the friction coefficient μ calculated for tests carried out on a laboratory stand, for various test conditions

Type of sleeve in the holder	Tests condition					
	No lubrication	Emulsion lubrication		Water lubrication		Pulse lubrication
		Flow rate Q 0.5 dm³/min	Flow rate Q 1,0 dm³/min	Flow rate Q 0.5 dm³/min	Flow rate Q 1,0 dm³/min	Flow rate Q 1,0 dm³/min
	The value of the friction coefficient μ					
Open with straight grooves	0.19	0.13	0.11	0.16	0.14	0.09
Open with helical groove	0.36	0.28	0.25	–	–	–
Closed smooth	0.21	0.08	0.07	0.14	0.12	0.06
Closed with helical groove	0.22	0.08	0.06	0.13	0.12	–

The view of the holder with the straight grooves sleeve and the closed smooth sleeve during the tests is shown in Fig. 5.20. The value of friction coefficient μ, calculated based on obtained torque value measurements, is presented in Table 5.3.

During the tests, significantly higher values of the resistance to rotation of the pick were recorded for the holder with a helical groove sleeve. Also, the higher values of the friction coefficient m were measured, which was probably caused by a faulty execution of the screw groove. This holder was eliminated after three series of tests from further trials.

The last tests were carried out with the lubrication of the holder by impulse supplying of the emulsion at a pressure of 1.5 MPa and a flow rate Q of 1.0 dm³/min, using an additional vane pump, a portion of about 5 cm³ of the enriched with 30% emulsion every 5 s. Improvement in results has been noticed.

The surface pollution with dust or sand negatively affects the tools rotation resistance. The resistance value for the holder with an open sleeve with straight grooves increased by 3–3.5 times and for the holder with a closed smooth sleeve more than 4 times.

The lubrication of dirty holders with emulsion significantly improved surfaces polluted with dust and sand. During the lubrication with the emulsion at a flow rate Q of 1 dm³/min and pressure p of 1 MPa, only after dozens of seconds, the value of resistance was lowered by a few times. The most favourable holder was a closed smooth sleeve and an open sleeve with straight grooves. The lubricant was flowing between the sleeve surface and the tool working part quite effectively washed out pollution. The value of rotation resistance measured after one minute of lubrication was only slightly higher than in the case of clean lubricating holders.

Based on obtained results, the holder with a closed smooth sleeve and the sleeve with straight grooves was selected for further tests on the stand for single tools examination—Fig. 4.2. Tests of mining with the use of holders lubricated with water and low-grade emulsion were performed. Standard conical picks with a carbide post diameter d of 22 mm were used for the tests.

The holders were lubricated by 1.5% emulsion at a pressure of about $p = 1.0$ MPa and flow rate Q of 1 dm^3/min and pure water at a pressure of about $p = 2.0$ and 3.0 MPa. Mining was conducted for a cutting depth $g = 9$ mm, cutting speed v_s of 3 m/s, cutting spacing $t_s = 12$ mm and total path of cutting at 2500 m at side deflection angle $\rho = 9°$. The concrete ring samples of compressive strength $\sigma_c = 65$ MPa were mined at the side surface.

The tests were also carried out with the holder's water lubrication with impulse supplying of the emulsion with the same previously given parameters. The number of tool rotations was measured for stable mining conditions during the tests. The wear of conical pick edges was also compared. One of the mining trials with conical pick mounted in holder with closed smooth sleeve is presented in Fig. 5.21.

During mining with tools mounted in holders lubricated by both water and emulsion, the more or less regular rotation of the tools was observed. The measured results are presented in Table 5.4. The highest number of rotations (over 10 rpm) was observed for the emulsion lubrication of the holder and the water lubrication of the holder with impulse supplying of the emulsion at side deflection angle appropriately $\rho = 6°$ and $9°$ [5].

It caused regular and slight wear of the edges of the tool. Moreover, both the working tool surface and cooperating surfaces of the sleeve and holder were after the sample mining clean without pollution. A view of one of the tool's edges and the surface of a holder with a closed sleeve, after tests of the rock sample cutting, with a water-lubricated holder is presented in Fig. 5.22. In the case of the holder with an

Fig. 5.21 A view of conical pick mounted in holder with closed smooth sleeve ready for test

Table 5.4 Number of revolutions of the conical pick, for various test conditions: type of lubricant fluid, the side deflection angle ρ, type of sleeve in the holder, fluid flow rate Q

Type of sleeve in the holder	Tests condition										
	No lubrication		Emulsion lubrication			Water lubrication				Pulse lubrication	
						p = 2 MPa		p = 3 MPa			
	Side deflection angle ρ, °										
	0	18	0	6	9	6	9	6	9	0	6
	Number of revolutions of the conical pick, n										
Open with straight grooves	0	0.50	1.25	5.25	9.25	2.25	3.25	2.25	5.50	2.75	7.50
Closed smooth	0	0.75	1.50	6.50	11.75	4.00	5.25	4.50	6.25	3.25	10.25

Fig. 5.22 View of the tool edge and surface of the water-lubricated holder with closed smooth sleeve, after tests of rock sample [1, 5, 7]

open sleeve with straight grooves, for the same mining conditions, the front surface of the holder and sleeve after the completion of the mining trials had traces of slight contamination (Fig. 5.23).

5.5 Field Tests of the New Solution of Lubricated Holder Solutions of Conical Picks

The positive results of the above-described tests allowed designing and manufacturing the unique mining head with lubricated holders for one of the most often used in underground mines light type roadheaders.

Fig. 5.23 View of the surface of the water-lubricated holder with open sleeve with straight grooves, after tests of rock sample [5]

The standard transverse mining head for this type of roadheaders, which was equipped with lubricated holders for conical picks, required the modernisation of its construction. Special channels in the mining head body and hydraulic aggregate for supplying lubricant agent for the conical picks holders were designed and manufactured [1, 3, 5].

Based on the results of laboratory tests, the holder with closed smooth sleeve and open sleeve with straight grooves were on the mining head body attached. The one cutter of mining head was equipped with holders with open sleeves with straight grooves and the second one with closed smooth sleeves. Through the canals drilled in the head body, the holders were supplied by the pure water under pressure p 1.5 MPa and flow rate Q 60 dm^3/min, circa 1 dm^3/min per holder.

A unique distributor used in the sector (internal) spraying system was attached to the mining head, which allowed to supply water under pressure only to tools in contact with the mined rock.

The view of the ready for field tests, modernised transverse mining head with lubricated holders, when the water supply was turned on, is presented in Fig. 5.24. Figure 5.25 presents a view of the holder with an open sleeve with straight grooves after the water supply is turned on.

Field tests with a mining head equipped with lubricated holders were carried out on a large concrete block with uniaxial compressive strength σ_c of about 60 MPa (Fig. 5.26). The block was mined at the same time (about 45 min) using a standard transverse mining head and a modernised mining head with lubricated holders. In the case of the mining head equipped with lubricated holders with open sleeves with straight grooves, a significantly more minor (at about 50%) and very even wear of pick edges was obtained, compared to standard mining head (Fig. 5.27). Deficient dust production was also observed. The surface of cutting tools and holders has been cleaned after the tests.

Fig. 5.24 View of the
modernised transverse
mining head of light type
roadheaders with water
supply turned on [5]

Fig. 5.25 View of the holder
with open sleeve with
straight grooves after the
water supply turned on [1, 5,
7]

Fig. 5.26 View of the
modernised transverse
mining head of light type
roadheaders, equipped with
lubricated holders, during
the large-size, concrete block
mining test [5]

Fig. 5.27 View of the pick
edge wear after 45 min of the
large-size, concrete block
mining: **a** mounted in the
holder with closed smooth
sleeve, **b** mounted in the
holders with open sleeve
with straight grooves [5]

Whereas, in the case of some holders with closed sleeves, a problem with the
correct assembly of the picks in the holders was noticed. These picks were fastened
inside the sleeve with the help of the expanding rings. This method did not fully
protect the pick from sliding out of the holder. In the case of several insufficiently
secured tools, due to their excessive sliding out of the holder, clearly greater wear of
their tip was noticed (Fig. 5.27a).

The results of the field tests were confirmed in the underground trials. These are
described in the next chapter.

5.6 Industrial Tests of the New Solution of Lubricated Holder Solutions of Conical Picks

The standard and modernised lubricated mining heads were tested during industrial
trials in one of the underground coal mines in Poland when a hard coal-rock face
was mined using the same light type of roadheaders [1, 5].

The comparative test of the standard mining head and the ones with lubricated
holders was performed for a more extended period (two working shifts per mining
head—total about 11 h). The view of the gallery face mined with the roadheader
equipped with a standard mining head and modernised lubricated mining head is
shown in Fig. 5.28. You can notice the significant, visible difference in the mining
process using both mining heads. In the lubricated mining head, the dustiness is
several times lower.

After the two mining shifts, the view of tested mining heads was also different.
Most of the tools mounted in lubricated holders did not show traces of wear or
pollution after tests. Figure 5.29 presents the tested mining heads after the trials.
The lubricated mining head was clean, whereas the standard one was completely
contaminated.

Additionally, over 50% of tools could not be used any longer and have to be
discarded on the standard head. The application of water-lubricated holders increased

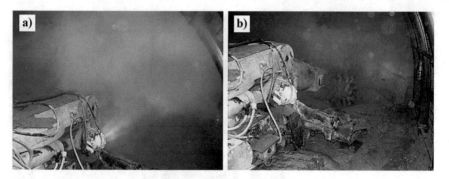

Fig. 5.28 View of the gallery face mined with the roadheader equipped with: **a** standard mining head, **b** modernised lubricated mining head [5]

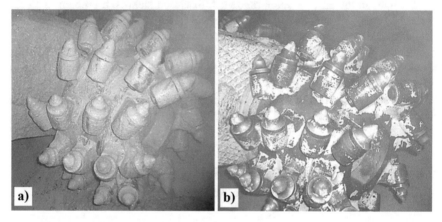

Fig. 5.29 View of the mining head after the underground trials: **a** standard mining head, **b** modernised lubricated mining head [1, 5]

the working life of picks almost twice. Comparing conical pick wear attached in the standard holder and lubricated holder with open sleeve with straight grooves is shown in Fig. 5.30.

The underground tests confirmed the defect of lubricated holders with closed smooth sleeves. The slipping out of the picks from the holders and their increased wear was observed. In some cases, the picks fell out of their holders during mining. A view of the damaged holder with a closed sleeve after the pick slipped out and hit the body of the roadheaders arm is shown in Fig. 5.31.

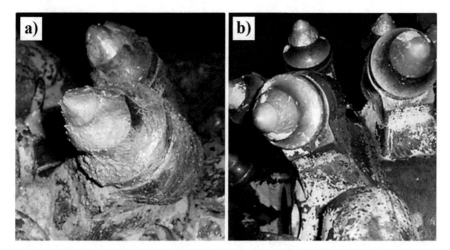

Fig. 5.30 View of the conical pick wear after the underground trials attached in: **a** standard holder, **b** lubricated holder with open sleeve with straight grooves [5]

Fig. 5.31 View of the damaged holder with a closed sleeve, after the pick slipped out

References

1. Kotwica, K., et al.: Collective Work—Selected Problems of Mining, Transport and Dressing of Highly Cohesive Rocks, Part 1, p. 147. AGH UST publisher (in Polish) (2016)
2. Kotwica, K., Kalukiewicz, A., Gospodarczyk, P.: Rotary Pick Holder for Mining Head, Especially Mining Combine Machines. Patent of the Polish Patent Office. No. PL 209806, Warszawa (2011)
3. Kotwica, K., Klich, A.: Machines and Equipment for Roadways and Tunnels Mining, p. 313. Instytut Techniki Górniczej KOMAG, Gliwice (2011) (in Polish)
4. Kotwica, K., Maziarz, M.: Impact of the mounting of tangential rotary tools on their proper operation. Arch. Min. Sci. **49**(1), 85–100 (2004)

5. Kotwica, K.: Application of Water Assistance in the Process of Mining Rock with Mining Tools. Monograph, p. 248. AGH UST Publisher. Kraków (in Polish)
6. Kotwica, K.: Effect of selected working conditions of cutting picks on their wear during the mining of hard rocks. Mech. Control **29**(3), 110–118 (2010)
7. Kotwica, K.: The influence of water assistance on the character and degree of wear of cutting tools applied in roadheaders. Arch. Min. Sci. **56**(3), 353–374 (2011)

Chapter 6
The Mining Head with Mini-Disk Tools with Complex Motion Trajectory

Abstract In the first part, a new conception of a mining head, in which the motion of tools will be forced and will cause mining of rock with tools along the complex trajectory, was described. This solution was developed in the Department of Mining, Dressing and Transport Machines, AGH University of Science and Technology, Cracow, based on an analysis of the world technique condition and results of own tests of rocks mining with asymmetrical disk tools of diameter D up to 160 mm. The mining method mode of this mining head solution allows the crossing of mining lines of individual disk tools and facilitates rock mining by breaking off rock furrows. The next part describes the laboratory trials to choose the most convenient geometric parameters of the symmetrical mini-disk tools and the way of the mining process. The tests were conducted on the unique laboratory stand in two phases—first only for a single plate with mini-disk tools with a straight-motion mining trajectory—in the horizontal or vertical plane. The concrete and natural rock samples were mined. The most favourable working conditions for obtaining the lowest load were determined. The geometric parameters of the symmetrical mini-disk were also selected while pressing the single mini-disk tools into the rock sample. The obtained results were checked on the laboratory stand for single mining tools testing during mining the concrete ring samples with a single plate with mini-disk tools with a complex-motion trajectory. In the next part, based on the selected working parameters of the new mining head solution, the mining head 3D model was designed, and a prototype of the head was manufactured. The new mining head was prepared for assembly on the arm of medium type roadheader. The field trials were performed on the large size concrete block to choose the most convenient working parameters of the prototype mining head—low load, extensive output graining, and small vibrations. Several parameters were checked, direction and number of rotations of the plates with disk tools and the mining head body, type of mini-disk tools, the direction of mining. The suggested solution of mining head with asymmetrical mini-disk tools of complex motion trajectory proved its usefulness while compact rocks mining and can be used as an alternative for existing mining heads of roadheaders. A selection of proper configuration of direction and number of rotations of the head body and plates with disk tools is necessary to obtain the most favourable parameters of the mining operation—the large size of output graining, the low load of drive engines and low vibration levels. The rotations of plates with disk tools should be in a clockwise direction, and the head body rotations in a counter-clockwise direction. The relation of the plates with disc tools rotations number to the head body rotations should be within the range of 3 to 3.3. It was stated that the most favourable direction of mining for this configuration of rotations was the horizontal one from right to left from the depth of 0 to maximum. The maximal depth value should not exceed 15–20 cm for the suggested head solution. The type of used mini-disk tools was significant—the used material and manufacturing technology. The most convenient results were obtained for the disk tools manufactured from tool steel type 36HNM—surface-hardened and tempered.

Practical application of the mechanical methods for hard rock mining is mainly connected with the gettability of a given rock. A significant criterion of the mechanical mining methods division is the direction of action of the main component of the tool force affecting the rock, which leads to its destruction. The direction can be tangent or normal to the surface of the mined rock.

Mining hard, strong mineable rocks, especially the ones that contain silicon on their chemical composition, its compounds or derivatives, or inclusions of effusive rocks with the application of methods based on cutting does not work. Such rocks require the use of high forces of cutting. Parallel application of the cutting pick and friction of the rock and the pick surface collaborated with it cause fast blunting (increased blunting increases friction force) and wear of the pick. New construction solutions of conical picks currently used on mining heads of roadheaders are equipped with the new generation of multilayer carbide inserts. However, the above-described mining conditions do not ensure adequate and economically justified rock centre mining. Also, applying internal and external water sparkling systems of cutting picks or high-pressure water jet assistance cannot fully provide even wear of the picks and their good durability.

The second most commonly applied method in the mechanical mining of compact rocks is mining with static pressure causing rock destruction by crushing realized by disk tools, usually symmetrical ones. The advantages and disadvantages of this method were in Chap. 3 described.

The asymmetrical disk tools have been used on mining heads of longwall shearers to increase the output of large size grains. Asymmetrical disc tools are applied in mechanical mining as crushing devices and chipping ones. Mining through the chipping with a disk tool uses typically for rocks with much smaller resistance against stretching than uniaxial compressive strength. The so-called back incision technique principle is mining a rock by cutting it off towards free space. This mining method was also in detail presented in Chap. 3.

A disk tool affects the rock tangentially to the surface of the mined body, similar to a cutting tool, but the difference is that it uses the disk rolling movement, which efficiently eliminates sliding friction in favour of rolling friction.

Application of the disk tools in that way lowers energy consumption and pressure force which allows constructing a mining machine of lower energy parameters and smaller size.

The machine this mining method was implemented allowed hard rock mining with lower energy consumption and significant output graining.

However, in the case of this method, there occur strongly changeable side forces on disk tool edges. The next drawback is a complex method of steering individual arms and very high reaction forces.

That is why the idea of the back incision technique was used for the development in the Department of Mining, Dressing and Transport Machines, AGH University of Science and Technology, Cracow, an innovative constructional solution of a mining head equipped with mini-asymmetrical disk tools that could apply in arm roadheaders. The course of implemented works is presented below.

6.1 The Conception of a New Solution of Mining Head with Mini-Disk Tools

Works connected with the development of new construction solutions of mining heads for arm roadheaders or mining methods allowing effective mining of hard mineable rocks have been conducted for years. In the Department of Mining, Dressing and Transport Machines, AGH University of Science and Technology, Cracow, an attempt was undertaken to solve a mining head equipped with asymmetrical mini-disk tools mounted in holders directly on the mining head body. Due to the high load of the disk tools and the low effectiveness of mining, the solution was given up.

Based on an analysis of the world technique condition and results of own tests of rocks mining with asymmetrical disk tools of diameter D up to 160 mm, the Department started tests to elaborate a new conception of a mining head, in which the motion of tools will be forced and will cause mining of rock with tools along the complex trajectory. This mining method mode allows the crossing of mining lines of individual disk tools and facilitates rock mining through breaking off rock furrows. It should decrease the energy consumption of the mining process. The disk tools were mounted on separate plates that could rotate on the mining head body and were propelled independently from it. Figure 6.1 presents the diagram of subsequent locations of the plate with disk tools during the head body rotation [4, 5].

It was assumed for simulation simplification that the plate with disk tools acts as a single disk tool. The following locations of the disk tool during the body rotation were determined, as shown in Fig. 6.1.

As the designed mining head will operate on an arm roadheader, its shape and dimensions should be close to the mining heads currently applied in the machines. In the back incision mining method, which was intended to be used in the developed conception of the mining head, mining requires the rotary motion of the disk tools.

Fig. 6.1 Diagram of
subsequent locations of the
disk tools during the mining
head body rotation [4]

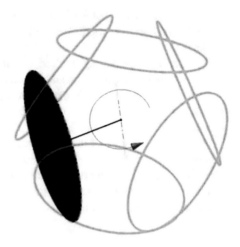

The rotation axes of the disk tools located on the plates have to move during work
concerning the mining head body to ensure this motion. Disks displacement during
the mining head operation can be obtained by locating the disk tools on a plate whose
rotary movement is enforced. The plates should be assembled to the side surface of
the mining head body, and their rotation axes should be perpendicular to the mining
head surface.

Based on these requirements, the solution of a mining head with asymmetrical
disk tools of a complex trajectory was elaborated. The disk tools were mounted on
three plates rotating concerning to the mining head body. Figure 6.2 presents the
scheme of this mining head solution [1, 4, 5].

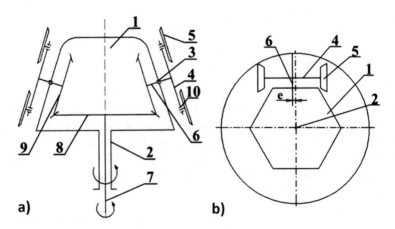

Fig. 6.2 Scheme of the mining head conception equipped with mini-disk tools of a complex motion
trajectory (description in the text) [1, 4, 5]

The developed mining head consists of an independently propelled body mounted in it and independently propelled three plates with asymmetrical disc tools. The mining head body 1 is propelled by an external drive shaft 2. In the head body, in seats 3, drive shafts 6 are mounted with the plates 4, on which in bearing seats 10 the disk tools 5 are mounted. The most favourable number of disk tools should be 6–8 pieces. The drive shafts 6 are propelled by an internal drive shaft 7, independent from the external drive shaft 2 and a set of bevel gears 8 and 9 or alternative ones.

In the conception of the new solution, it is crucial to ensure dislocation of the drive shafts axes 6 of the plates 4 with mining tools 5 by the value e so that they do not cross the axis of the drive shaft 2 of the head body 1. The eccentric location of the plates axes with disk tools should enable their easy slotting into the rock massive at the unit motion, both vertical and horizontal. This mining head solution has been submitted to the Polish Patent Office and is covered by patent protection No. PL 209326 [6].

The trajectory of disks motion can change depending on the plane's inclination in which the plate rotates concerning the body rotation axis. Figure 6.3 shows the result of a simulation of disk tools trajectories, performed for four angles of the plate plane inclination 0°, 10°, 20° and 30°. The simulation was performed not just for one plate but for all three mounted on the head body. It was stated that at higher values of inclination of the plane in which the plate rotates concerning the body rotation axis, the mining zone moves towards the unit face, which facilitates breaking the rock chips.

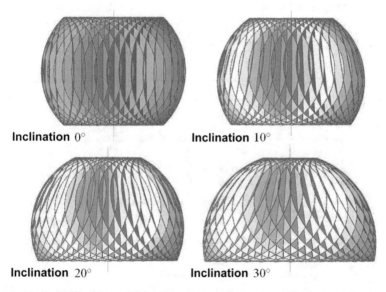

Inclination 0° Inclination 10°

Inclination 20° Inclination 30°

Fig. 6.3 Result of a simulation of disk edges motion trajectory depending on the inclination angle of the plane in which the plate rotates concerning the rotation axis of mining head body [4]

Fig. 6.4 Result of simulation of the displacement of the plates with disk tools, depending on the plate plane inclination angle and eccentric displacement e of the plates rotation axis against head body axis [4]

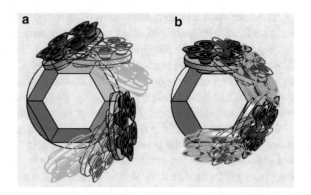

Figure 6.4 presents an exemplary result of a simulation of displacement of plates with disk tools during a rotation of the mining head body. Such a simulation was also performed to check the mining depth with the single disk tools. For the simulation, the eccentric e value by which the axes of the plates with disk tools were moved towards the body rotation was changed. Also, a change of inclination angle of the plane in which the plate rotates concerning the body rotation axis was made.

Figure 6.4b presents trajectories of disk tools obtained for the axis of plates rotation perpendicular to the body surface and crossing in the axis of the body rotation. Total mining depths did not exceed a dozen millimetres, and the plates with disk tools abrasive rubbed the mined rock. After increasing the value of inclination of the plates rotation axes and moving them eccentrically towards the head body rotation, several times deeper mining was obtained without plates rubbing the rock. A view of such a simulation is presented in Fig. 6.5a. The most favourable parameters from

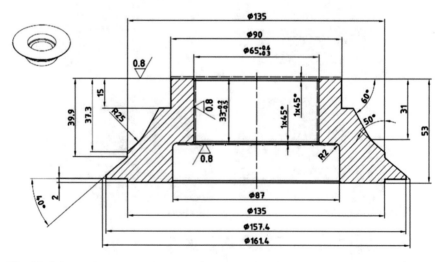

Fig. 6.5 Scheme of an asymmetrical mini-disk tool used in the first phase of the research [1]

the performed simulation were obtained for the plate axis inclination angle of 15°
and eccentric displacement towards the head body rotation axis by about 20 mm.

For the conception of the new mining head with mini-disk tools of motion trajectory, their values were accepted as the most beneficial ones in the next part of the
research. This solution also allows the location of the gear inside the body to keep
a favourable mining head shape. The shape and size of the mining head should be
close to solutions of units presently used on the roadheaders.

The prototype solution of the new mining head necessary was also to determine
values of many other parameters such as e.g. a number and rotation direction of plates
with tools, required value of torques for plates and the head body, and dimensions
of disk tools.

The theoretical estimating of these values was very complicated or, in some cases,
even impossible. The empirical tests were proposed using a specially constructed
laboratory stand and the stand earlier described in Sect. 4.2. Results of these works
are presented below.

6.2 Laboratory Tests of the New Mining Head Solution and Mini-Disk Tools

In the first part of the laboratory tests of the mining process with asymmetrical mini-
disk tools, the tool solution with diameter $D = 160$ mm and disk edge angle $\beta = 40°$
was planned for use. The scheme of this disk tool is shown in Fig. 6.5. For constant
geometric parameters of the disk tool, it was necessary to determine the required
operating parameters for the developed mining head solution [1].

The maximal diameter of the plate for disk tool assembly was defined at 450 mm,
the number of disk tools at 6 pieces and their distribution on a diameter equal to
350 mm. The required torque for the plate drive was calculated at a minimum of
3000 Nm based on the previously conducted tests on another laboratory stand. The
number of rotations of the plate should reach up to 60 rpm, their steering should be
smooth, and there must be a possibility to change the direction of the rotations. Most
favourite plate drives were a hydraulic engine with toothed gear for these conditions.
The hydraulic engine of OMTS 250 type and integrated planetary gear of RR5/OMC
type made by the Sauer Danfoss company was proposed.

For tests, the plate with disk tools must move in three mutually perpendicular
directions at changing. The velocity of the movement should be smoothly controlled.

6.2.1 Laboratory Tests of a Single Plate with Mini-Disk Tools with A straight-Motion Trajectory

Based on all the above-presented requirements, a solution of laboratory stand was designed and manufactured to test mini-disk tools. The spatial model of this stand is presented in Fig. 6.6. The lab stand was developed to tests a single plate with mini-disk tools with a straight-motion trajectory—in the horizontal or vertical plane [3, 4].

The main element of the stand is the welded frame. At the front part of the frame is attached a construction of decking for an artificial rock sample execution of dimensions 1650 mm × 1000 m × 1200 and volume of almost 2.0 m³. Between the decking and the rear part of the frame were assembled two pipe slides, on which the swivel base with manipulator and research head with disk tools. The rotary plate was moved forwards and backwards with a hydraulic actuator.

The plate with disk tools was mounted to the arm and propelled by a described drive unit installed on the other side of the arm. The arm with the plate with disk tools was rotated horizontally by rotation of the base of the rotary plate, using the hydraulic actuator. A hydraulic actuator mounted between the base and the arm construction raised and lowered the arm. For the proposed constructional solution, kinematic parameters of the plate with disk tools were determined. They are presented on the scheme in Fig. 6.7. The plate featured an ability to move forward and backward at the length of 1000 mm, turn by 32° to the left or right, whereas in vertical movement, it can be lowered by 17° and raised by 12°. It allowed mining the rock sample at its whole face surface at a depth comparable to the length of the sample.

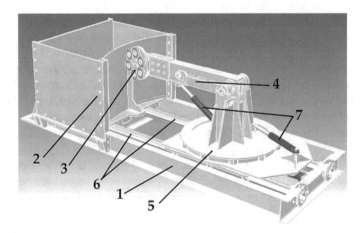

Fig. 6.6 The spatial model of a laboratory stand for testing the mining process of compact rocks with a single plate equipped with asymmetrical mini-disk tools of straight-motion trajectory: 1—mainframe, 2—artificial rock sample decking, 3—plate with disk tools and drive unit, 4—jointed arm, 5—rotary plate, 6—two pipe slides, 7—two hydraulic actuators according to [4]

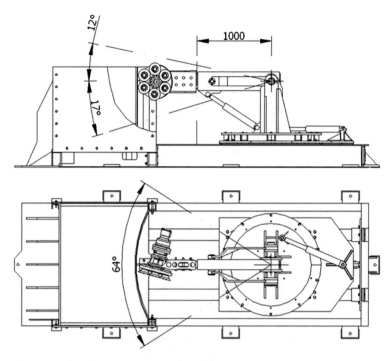

Fig. 6.7 Scheme of the laboratory stand for testing the mining process of compact rocks with disc tools of complex trajectory with marked kinematic abilities [4]

The elaborated spatial model of the laboratory stand was checked by load simulations using the Finite Elements Method in the Ansys Design Space pack. Due to limited possibilities, the simulation was static without modelling the mined rock centre. For calculation was assumed that one of the disk tools is fixed at a depth of 15 mm in a brittle-elastic centre, whereas hydraulic actuators generated forces constraints of maximal possible values. A simultaneous combination of three possible movements on the stand—forward movement, rising and arm rotation was also considered. The material planned for the stand implementation was an alloy steel 18G2A for the most extreme work conditions. In the simulation results, no symptoms of increased load of the material were observed.

Based on the developed spatial model, the laboratory stand was manufactured. The stand was supplied and controlled from a unique hydraulic aggregate allowing the smooth setting of the value of pressure and flow rate and change of the plate with the disk tools rotation direction. The aggregate parameters enabled to obtain the number of rotations of the plate with disk tools within the range from 0 to 65 rpm. Whereas, for used hydraulic actuators, the following velocity values of the plate with disk tools were obtained: forward and backward from 0 to 7.2 cm/s, vertically from 0 to 8.9 cm/s, horizontally from 0 to 4 cm/s. All the values were within the required range for a research stand.

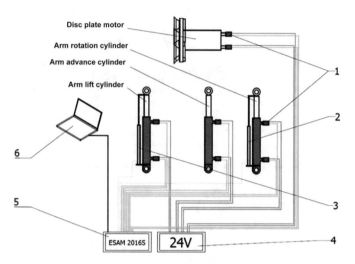

Fig. 6.8 Scheme of the measuring system and arrangement of measuring sensors: 1—pressure transducer MBS32, 2—movement sensor PT300, 3—movement sensor PT500, 4—feeder, 5—measurement card ESAM 2016S, 6—measurement computer [4]

Apart from the observations of the mining process with disk tools of complex trajectory, it was also planned to determine the values of approximate moments affecting the plate with disk tools and the forces from individual movements of the plate. In the system supplying individual hydraulic receivers were installed pressure transducers made by Danfoss type MBS32, registering pressure values at both the supply and downflow. Additionally, the hydraulic actuators were equipped with induction sensors of movement PT, allowing defining velocity of the plate with disk tools. The scheme of the measuring system and arrangement of measuring sensors is presented in Fig. 6.8 [4].

In the first part, for the mini-disk tool presented above, it was accepted to perform tests comprising at least four changing parameters: the direction and number of rotations of the plate with disk tools (forward and backward rotations and at least three ranges of the number of rotations—40, 50, and 60 rpm), the mining spacing t (from 10 to 40 mm), the cutting depth (at least three values of cutting depth g from 6 to 20 mm). The above-listed parameters were assumed to perform cuts, moving the plate horizontally relative to the rock sample and then vertically.

A diagram of the mining process of the rock in both planes is presented in Figs. 6.9 and 6.10. Depending on horizontal mining direction (left or right) and vertical mining from the left or right rock sample edge, the asymmetrical disk tool attacked the rock sample with its flat or inclined surface of the edge. The tests aimed to assess the mining process efficiency with the plate with disk tools depending on the movement mode and direction of the disk plate rotations [4].

The first tests of sample mining in a horizontal plane were conducted only on a concrete block of uniaxial compressive strength σ_c of 26 MPa. The realisation of the research started with pilot tests during which the sample face was adequately

Fig. 6.9 Diagram of
horizontal mining of the
samples: **a** to the left, **b** to
the right [4]

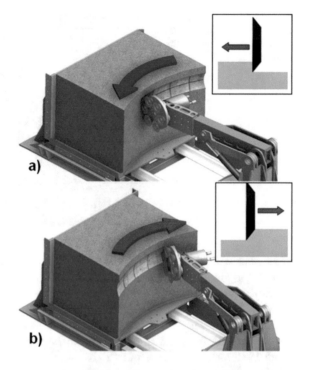

Fig. 6.10 Diagram of
vertical mining of the
sample: **a** from the left edge,
b from the right edge [4]

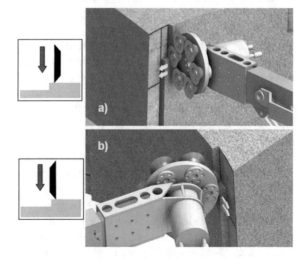

equalised. The cutting depth g was not bigger than 10 mm, the number of the plate
with the disk tools rotations was in the range of 40–60 rpm. At speed about 60 rpm a
significantly lower vibration of the construction was observed, which considerably
improved work conditions of the plate with disk tools. Comparing the rotary direction

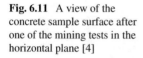

Fig. 6.11 A view of the concrete sample surface after one of the mining tests in the horizontal plane [4]

of the plate with disk tools could be stated that forward rotation of the plate allowed minimising vibrations of the construction stand. A view of one of the mining tests in the horizontal plane is shown in Fig. 6.11 [4].

After the sample surface alignment, the test of concrete sample mining in the horizontal plane was started. The cutting depths values were in the range from 6 to 15 mm were assumed, and sample mining on the entire width of the sample surface—about 1630 mm, from the left and right sample edge alternately, with the forward rotation of the plate with the constant number of rotation 60 rpm.

During the concrete block mining from the left to the right side, a significantly lower dynamics of the mining process was observed, resulting from the way the artificial sample attacked the flat surface of the disk tool. A view of courses of measured component forces values (pressure force Pd, tangential force Ps and side force Pb) during mining to the left and right at cutting depth g 10 mm was presented in Fig. 6.12.

The main part of the mining tests with the plate with disk tools in a vertical plane was carried out on the concrete block. Some mining tests were repeated on natural rock samples embedded in a concrete block. Sandstone block of uniaxial compressive strength $\sigma_c = 73$ MPa and a granite block of $\sigma_c = 253$ MPa were used for the tests. After embedding in the concrete sample, these blocks, shown in Fig. 6.13, were ready to test.

Tests of vertical mining started with mining the concrete block. The first cut was performed from the left edge of the concrete sample, at the cutting depth $g = 10$ mm and cutting spacing $t = 10$ mm. Then, half-open cuts were done for cutting spacing $t = 20$ mm, 30 mm and 40 mm. In the next series of tests, the cutting depth was increased to 15 mm, repeating values of applied cutting spacings from the previous series. For both series, it was noticed that for the cutting spacing $t = 40$ mm, full spalling of the cut groove does not occur, and at the depth $g = 15$ mm, strong reactions occur, leading to the construction vibrations that disturb the mining process.

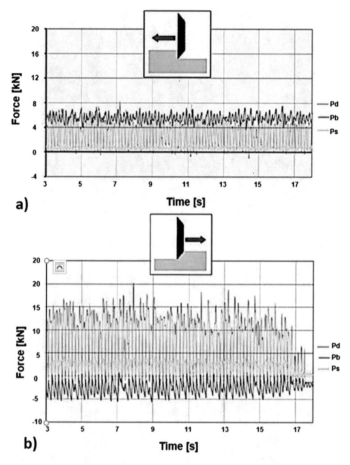

Fig. 6.12 Courses of values of pressure Pd, cutting Ps and side Pb forces during mining at cutting depth of g 10 mm to: **a** the left side, **b** the right side [4]

The next part of the tests was performed from the right edge of the sample when the disk tool attacked the mined sample with the inclined surface of the disk edge. The tests were carried out at cutting depth of $g = 15$ and 20 mm and cutting spacing $t = 20$, 30 and 40 mm. In every mining case, full spalling of the cut groove has been noted. Therefore, additional tests were performed for cutting depth $g = 25$ and cutting spacings $t = 20$, 30 and 60 mm. For the first two cutting spacings, full spalling of the cut groove occurred. The cut groove was not fully spalled for the cutting spacing $t = 60$ mm. The average values of component force measured during left and right vertical mining of the concrete block at the cutting depth $g = 15$ mm were presented in Fig. 6.14.

Due to the relatively small size of the sandstone and granite samples, it was decided to apply a constant cutting depth and perform cuts only from the right side during their vertical mining. The cutting depth for sandstone sample $g = 40$ mm at

Fig. 6.13 A view of sandstone (1) and granite (2) blocks embedded in the concrete sample [4]

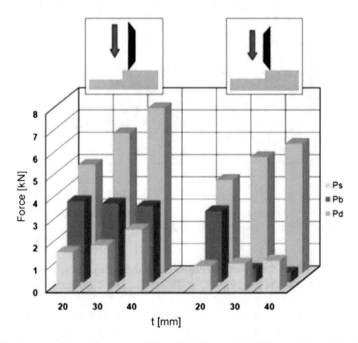

Fig. 6.14 Average values of the pressure Pd, cutting Ps and side Pb forces during mining from the left and right side of the block at cutting depth g 15 mm [4]

the cutting spacings $t = 30, 40, 50$ and 60 mm, respectively, was assumed for the granite sample $g = 25$ mm at $t = 10, 20$ and 30 mm.

Fig. 6.15 A view of the sandstone block surface after completing the mining tests in the vertical plane [4]

In comparison to tests of concrete sample mining, a significant decrease in the mining process dynamics was noticed. After sandstone and granite sample mining, the obtained surface was very even, which meant a full spalling of cut grooves (Fig. 6.15).

Vertical mining from the right edge sample allowed higher values of the cutting depth and cutting spacing at full spalling of the cut groove. The test results show the influences of mining parameters on values of component mining forces. The value of pressure force Pd reached the highest values significantly. An increase in the pressure force value was caused by the increased cutting depth and spacing [4].

The value of tangential force Ps changed mainly because of cutting depth. The change of cutting spacing did not affect the tangential force value (especially for sandstone). In the case of the value of side force Pb the significant influence of mining parameters was observed. In the case of the concrete sample mining, the highest values of the side forces were connected with a value of cutting spacing when full spalling of the cut groove was not obtained. Figure 6.16 shows the average values of mining forces for the mining tests of concrete and natural rocks samples obtained for the highest values of cutting depth g and cutting spacing t [4].

The visual inspection of the obtained output graining was also performed. As a result of vertical mining, the dominant fraction are grains several times bigger than grains obtained during horizontal mining (Fig. 6.17). It mainly refers to when mining took place from the right side of the block, and the disk tool attacked the rock with an inclined edge surface.

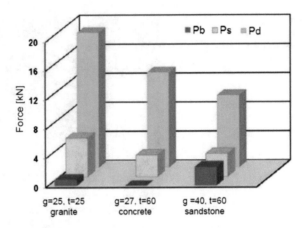

Fig. 6.16 Average values of mining forces for maximal values of cutting depth and spacing during mining different rock samples [4]

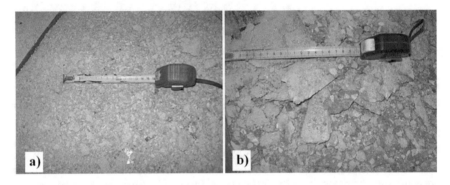

Fig. 6.17 Graining of the output obtained during horizontal (**a**) and vertical (**b**) mining [4]

6.2.2   Laboratory Tests of Pressing the Single Mini-Disk Tools into the Rock Sample

Previous researches on mining with asymmetrical disk tools, which were carried out on a specially designed stand at the Department of Mining, Processing and Transport Machines, AGH UST, focused on the tests of the single plate with disk tools.

Based on the construction and movement capabilities of the described stand, the modernisation of the stand was carried out to be used for static tests of pressing a single disk tool into the rock sample [1, 7].

For this purpose, a special head with the holder has been developed to mount a disk tool and directly measure the pressure force when pressing the tool into the rock sample. The model and view of this measuring solution are shown in Fig. 6.18. The device consists of a mounting frame 1, to which the measuring head 2 is screwed. In frame 1, a disk tool holder 3 is mounted together with an asymmetrical disk tool

Fig. 6.18 A spatial model and a view of a new system for measuring the pressure force of a disk tool: 1—clamping frame, 2—force sensor, 3—disk tool holder, 4—disk tool [1]

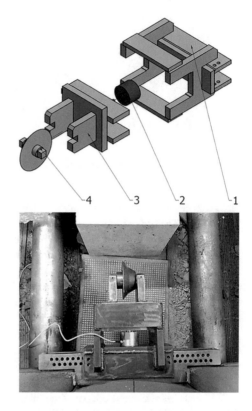

4. The frame has holes, which enable bolt fastening to the pitch bar, welded to the sliding frame of the rotary table.

The hydraulic actuator of the laboratory stand's feed system pressed the disk tool into the rock sample in the form of a cube with a side of 400 mm. The sample was attached to the stand structure with a unique frame.

The head for measuring the pressure force of the disk tool is equipped with a strain gauge force sensor, which allows measuring the force in the range of $0 \div 200$ kN. A significant advantage of the system is the possibility of simultaneous measurement of the displacement of the disk tool concerning the mined sample. An induction displacement sensor type CL460 performed displacement measurement with a measuring range of $0 \div 300$ mm.

The purpose of the research was to identify the impact of geometric parameters of asymmetrical mini-disk tools on the values of loads generated while pushing a disk tool into a rock sample. Static tests consisted of pushing the disk tool into a rock sample at a given mining spacing from the sample edge until a piece of rock breaks off. The following geometric and process parameters were adopted as independent variables for the tests: the diameter D and an edge angle β of the asymmetrical mini-disk tool, mining spacing t and the uniaxial compressive strength σ_c of the sample. The tests were carried out for disk tools with the diameter $D = 150$, 160 and 170 mm

Fig. 6.19 The set of asymmetrical mini-disk tools used in the second phase of the research [1]

and edge angles $\beta = 35°$, $40°$, and $45°$. Figure 6.19 shows the view of the disk tools set used for the tests [1, 7].

The construction of the disk tool holder permitted changing the disk's position relative to the sample and thus allowed setting the correct cut spacing. The disk tool's *Pd* pressure force value and displacement were measured and recorded during the research. Research on the disk mining process was performed for concrete samples with uniaxial compressive strength σ_c of 25 MPa and sandstone samples with uniaxial compressive strength σ_c of 79 MPa.

Each of the mining tests consisted of setting an appropriate mining spacing by the appropriate aligning and fixing of the holder together with the disk tool. In the next step, the disk penetration was made close to the edge of the rectangular rock sample until the rock piece was broken off from the sample.

The tests were carried out for three mining spacing values $t = 15, 25$ and 35 mm. After each of the tests, the extent of the destruction zone of the sample was measured (Fig. 6.20). Performing another attempt required moving the holder towards the opposite, intact edge of the sample and rotating the disk tool so that the chipping was obtained by the inclined surface of the disk edge [1, 7].

Each of the trials for a given configuration of the geometric disk parameters and the determined mining spacing was repeated at least three times. Most tests were completed by breaking off a piece of rock whose dimensions (height, depth) were multiple times higher than the assumed mining depth. Exemplary results of pressing the disk tool 150 mm in diameter and edge angle $\beta = 40°$ into the sandstone sample are presented in Fig. 6.21.

Figure 6.22 presented an example of graphs of the impact of selected geometric parameters of asymmetrical mini-disk tools and mining spacing on the value of pressure force Pd. The average values of maximum pressure forces were calculated based on at least three measurements. Based on the research results, it can be stated that the increase in the value of individual independent variables caused an increase in the generated pressure force. The dynamics of the pressure force values changes depending on the variable adopted as the input parameter for the tests.

The most significant increases in the pressing force value were observed concerning the cutting spacing t (Fig. 6.22a, d). The change of such parameters as the diameter of the disk tool D (Fig. 6.22b) or the angle of the disk edge β (Fig. 6.22c)

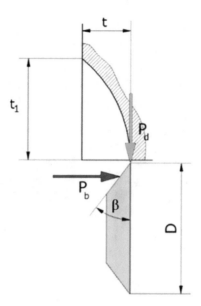

Fig. 6.20 Scheme of measured parameters during mining tests [7]

caused a minor increase of the pressing force than the mining spacing. The influence of the sample compressive strength σ_c on the value of the pressing force of the disk tool was also checked. Based on the results of the tests, it can be concluded that the increase in the compressive strength of the sample caused a proportional increase in the registered value of the pressure force.

6.2.3 Laboratory Tests of a Single Plate with Mini-Disk Tools with A Complex-Motion Trajectory

For the solutions of disk tools presented in Fig. 6.7, the last part of tests was proposed to perform on the modernized laboratory stand for single mining tools testing (Fig. 4.2). The laboratory stand was adapted for assembly of the plate with disk tools and drive unit—the same one used in the stand described in Fig. 6.6. It was necessary to develop a unique mounting of the plate assembly with the drive attached to the support. It allowed the mining of the ring rock sample on the side surface with its complex rotary movement and the plate with disk tools [1].

The unique three-axis sensor of forces and moments type CL 16 was attached between the plate with disk tools and the drive unit. The spatial model of the laboratory solution stand presents in Fig. 6.23. Figure 6.24 shows the view of the modernized laboratory stand, ready to test [1].

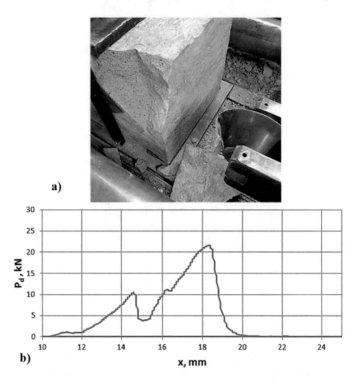

Fig. 6.21 A view of the effect of pressing the disk tool into the artificial sample—diameter D = 150 mm, edge angle β = 40°, mining spacing t = 35 mm: **a** view of a chipping fragment of a rock block, **b** a course of pressure force value according to [1]

The initial tests were carried out for determining the direction and number of revolutions of the rock sample and the plate with disk tools. The disk tools with a diameter D of 160 *mm* and an edge angle β of 40° were used. Convergent and opposite rotation directions were checked for the plate with disk tools constant number of revolutions at about 60 rpm and table rotation with the ring sample at about 20 rpm. The sample was mined from the top with the mining $g = 15$ mm depth.

The disk tools penetrated relatively gently into the sample in the convergent direction of rotation, breaking off the concrete fragments. In the opposite direction of the rotation, the sample and the disk tool collided violently, generating significant dynamic loads and vibrations of the laboratory stand. The convergent direction of the sample and plate rotations was selected for further mining trials.

Four depths values of mining were selected for further tests, $g = 15, 25, 35$ and 45 mm. The first mining trial view at a mining depth of 15 mm is shown in Fig. 6.25a. Figure 6.25b shows the result of a ring sample mining test along with its entire height, with a mining depth of 30 mm. During the mining, cyclic grooves on the surface of the sample, related to the regularity of successive chipping of the concrete material by disk tools, were observed. There were no significant signs of disk tools edge wear

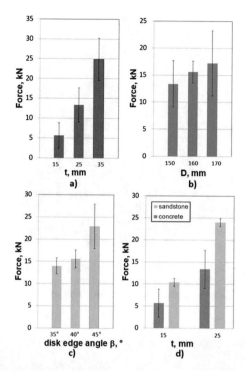

Fig. 6.22 Average values of maximum pressure forces for: **a** disk tool diameter D = 160 mm and sample compressive strength σ_c = 25 MPa, **b** mining spacing t = 25 mm, edge angle β = 40° and sample compressive strength σ_c = 25 MPa, **c** disk tool diameter D = 160 mm, mining spacing t = 25 mm and sample compressive strength σ_c = 25 MPa, **d** disk tool diameter D = 150 mm and sample compressive strength σ_c = 25 and 79 MPa [7]

observed. Depending on the mining depth, the large granulation output was obtained from several to over 100 mm (Fig. 6.26) [1].

The test results confirmed that using asymmetrical mini-disk tools allows effective rock mining, with a large grain size of the obtained output. The selection of mining parameters is for this method essential. Based on the test results, the following parameters were chosen [7]:

- the vertical direction of mining,
- the convergent directions of the plate and sample rotation,
- the constant number of revolutions of the plate with disk tools at about 60 rpm,
- the constant number of revolutions of the table with the ring sample at about 20 rpm,
- the mode of mining with the disk edge inclined surface from the side of the free surface of the rock.

In the last part of the trials, the influence of geometrical parameters of disk tools on their wear was checked. It was assumed that each set of disk tools (Fig. 6.19) would

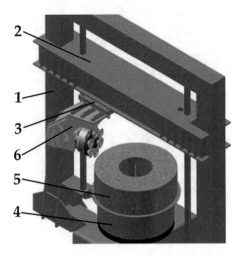

Fig. 6.23 The spatial model of a laboratory stand for testing mining process with asymmetrical mini-disk tools of complex-motion trajectory: 1—mainframe, 2—traverse, 3—support, 4—rotary table, 5—mined rock sample, 6—a unique device for the fixing of a plate with mini-disk tools and drive unit [7]

Fig. 6.24 A view of the modernized laboratory stand ready for the tests [1]

mine a constant and comparable volume of output equal to about 2/3 of the volume of the mined concrete sample. Before and after the mining process, the state of the tools was compared using the 3D scanning technique by the HDI Radwance scanner from LMI Technology. It allowed the comparison of the size and type of wear of used types of disk tools [1].

Six concrete ring samples were made for testing, with the compressive strength σ_c of 80 MPa. The ring concrete sample was successively mined with layers from the top to down, from outside to centre. The mining depth $g = 25$ mm and the advance

Fig. 6.25 A view of the sample: **a** during mining at the mining depth of 15 mm, **b** the surface after mining at the mining depth of 30 mm

Fig. 6.26 A view of the obtained output grains with size of over 100 mm

speed along the mined sample surface equal to 30 mm per sample revolution were constant for each trial.

Based on the obtained results, it can be stated that the disk tools with the largest diameter showed the most extensive edge wear and the decreasing of the outside diameter. In the case of the disk tool with 150 mm diameter, the diameter value loss was much lower.

When comparing disk tools with variable edge geometry, it can be observed that the disk tools with the most minor edge angle value $\beta = 35°$ wore the fastest, while the minor wear showed the disk tools with the edge angle value $\beta = 45°$. An exciting observation was that all the disk tools had a rounded edge after the mining. The radius of the disk rounding increased from the initial 2 mm to an average of 7.5 mm. In addition, for all disk tools after the mining, the angle of the edge β increased up to about 90°. Tables 6.1 and 6.2 present the results of this research [1].

Table 6.1 Changing the geometric parameters of the disk tools after the tests (disk tools of different diameters D)

Type of the disk tool	D = 150 mm, β = 40°	D = 160 mm, β = 40°	D = 170 mm, β = 40°
Changing the outer diameter of the disk tool (mm)	−2.5	−4.5	−9.83
Edge angle value (°)	86	91	89
Edge radius value (mm)	5.8	7.6	8.2

Table 6.2 Changing the geometric parameters of the disk tools after the tests (disk tools of different edge angle β)

Type of the disk tool	D = 160 mm, β = 35°	D = 160 mm, β = 40°	D = 160 mm, β = 45°°
Changing the outer diameter of the disk tool (mm)	−5.3	−4.5	−2.77
Edge angle value (°)	89	91	88
Edge radius value (mm)	8.3	7.6	6.9

After the mining process, it was observed that on the edge of the disc tools with edge angles $β$ of 35° and 40°, local damages and breakouts occurred. The view of this type of damage for the disk tools with diameter $D = 160$ mm and edge angle $β = 35°$ is shown in Fig. 6.27. This phenomenon was not observed for the disc tools with edge angle $β = 45°$.

Fig. 6.27 A view of local damages and breakouts occurred for the disk tool with diameter $D = 160$ mm and edge angle $β = 35°$ after the tests

6.3 Developing and Manufacturing of New Solution of the Mining Head with Mini-Disk Tools of Complex-Motion Trajectory

Based on the obtained laboratory tests results, for further designing works over the prototype mining head with disk tools of the complex trajectory, the following requirements were selected [3, 5]:

- Diameter of disk tools $D = 160$ mm,
- Disk tool edge angle $\beta = 40$–$45°$,
- Diameter of plates with disk tools 700 mm $\leq$ Du $\leq$ 850 mm,
- Diameter of disk tools distribution on the plates 550 mm $\leq$ Dn $\leq$ 650 mm,
- Number of plates with disk tools—3,
- Number of disk tools on the plates $I_d \leq 8$ pieces,
- Revolutions number of plates with disk tools $60 \leq n_p \leq 100$ rpm,
- Revolutions number of the mining head body on which the plates are mounted $20 \leq n_k \leq 30$ rpm,
- Torque value of the one plate with disk tools $M_t \approx$ minimum 4000 Nm, suggested 5000 Nm,
- Total torque value to propel the plates with disc tools—minimum 8000 Nm, suggested 10,000 Nm.

The above-presented values were estimated in comparison to the results obtained in laboratory stand tests (increased diameter of the plates with disc tools up to 50%, expected higher compressive strength of the mined rock at the level over 100 MPa, and target mining depth at minimum 25 mm) and the fact that during mining there are at least two plates with disk tools in contact with the rock body at the same time.

Based on the above-presented requirements, the design and spatial model of the new head solution with disc tools of complex trajectory have been elaborated. The works were performed with the REMAG Ltd. company—the Polish producer of light and medium roadheaders. The new mining head solution was executed for a medium type roadheader of 52 tonnes manufactured by REMAG. Based on the analysis of the parameter of mining heads used on the roadheader, the length of the new head solution should not exceed 1750 mm, the diameter 860 mm and weight 5 tonnes [3, 4].

In the assumption of the mining head design, the plates' drives with disk tools and head body were to be independent. The separate drives for both motions should enable independent control of rotational speed value. The head body was propelled by a hollow shaft, whereas an external shaft realised the drive of the plates with disk tools through a gear placed inside the body. The scheme of drive transmission is shown in Fig. 6.28 [3, 4].

The gear solution in the new mining head allows displacement of the plates rotation axle concerning the axle of the whole unit rotation. In this option, the driving wheel 2 propelled by shaft 1 cooperates with wheels 3 mounted on the shafts with wheels 4 from which the torque is transmitted to conic wheels 5 propelling individual plates.

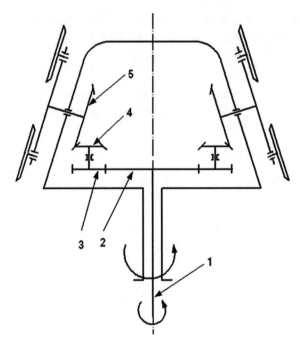

Fig. 6.28 Optional solution of inner double bevel cylindrical gear for drive transmission on plates with disk tools [3, 4]

In the developed initial model of the new mining head, three plates with disk tools on its body were mounted. The head body had an ability of independent rotation concerning plates with disk tools. The kinematic capabilities of the new head solution are presented in Fig. 6.29.

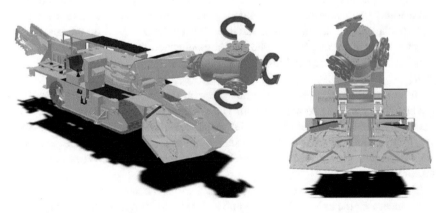

Fig. 6.29 Kinematic capabilities of the new head solution mounted on the medium type roadheader [3, 4]

Based on the above diagram, the technical project of a mining head with asymmetrical disk tools of complex trajectory was elaborated. The spatial model and the view of the new mining head prototype are shown in Fig. 6.30. For the initial solution was chosen an electrical engine of 150 kW power mounted in the roadheader arm that drives the plates with disk tools. The drive of the head body was realised by two hydraulic engines HS 0.8.

The engine propelling the plates should not exceed 50 kW, and the engine revolutions should be no higher than 500 rpm. For these parameters, data and maximal rotations of plates reaching 200 rpm, a two-step cylindrical-bevel gear of summary ration about 1:2 was elaborated.

The gear transmitted rotations onto the head body from two engines HS 0.8 of power of 2 × 80 kW, at maximal engines rotation 250 rpm. The external cylindrical

Fig. 6.30 Model and the view of the new mining head solution with disk tools of complex trajectory: 1—main gear, 2—auxiliary gear, 3—disk plate, 4—input shaft, 5—central gear wheel, 6—orbital gear wheel, 7—pinion, 8—face gear, 9—output shaft, 10—support, 11—connector, 12, 13—bearings, 14—body gear, 15—hydraulic engine according to [3, 4]

Fig. 6.31 A view of the new
mining head solution with
asymmetrical disk tools
during one of the field tests
[1]

gear of ratio 1:6 was elaborated for planned head body rotations reaching 30 rpm [4].

Based on the elaborated model, a prototype of the head was manufactured. After assembly on the roadheader, the new mining head during the preliminary field tests is shown in Fig. 6.31.

6.4   Field Tests of the New Solution of the Mining Head with Mini-Disk Tools of Complex-Motion Trajectory

Preliminary field tests of the mining head with asymmetrical disk tools of complex trajectory were conducted on the special test stand with a large-size concrete block (6 m width × 4 m height × 5 m depth) of uniaxial compressive strength of the range 40 MPa in the bottom up to 76 MPa in the top [4, 5, 7].

During the first stage of preliminary tests, a change of the head body rotations value was planned within 10–30 rpm and for the plates with disk tools from 20 to 200 rpm. The rotation direction was also changed. The cutting depth of the head did not exceed 50–100 mm. Mining effectiveness, the grain size of the output and the level of the tools wear were checked.

The trials started at the rotations of plates with disk tools in a clockwise direction and the head body rotations in a counter-clockwise direction. Disk tools were mounted on the plates with the flat side out. The new mining head solution worked for the above parameters without serious reservations. At lower values of plates rotations, very large graining was obtained. However, it was accompanied by high vibrations of the head.

An increase of rotations of the plates with disk tools caused lowering the vibrations but increased dustiness and at maximal rotations, there even occurred sparking at the contact of the tools with the rock. The best effects of the mining—significant graining of the output, low engines load and little vibrations were obtained for the head body rotations of ca 20 rpm at the plates rotations value of ca 60 rpm. A view of the obtained winning and an atypical surface of the mined concrete sample (intersecting of the cuts) is shown in Fig. 6.32. No visible symptoms of the disk tools edge wear were noticed. The change of direction of the head body or plates with disk tools rotation onto the opposite one had a highly negative influence on the engines load and disk tools and the plates themselves wear (Fig. 6.33a).

The next part of mining tests of the concrete block for different mining directions was performed, at the rotations number of plates with disc tools at ca, 70 rpm and clockwise direction, and rotations of the head body ca 20 rpm and opposite direction. The values were changed within a narrow range, keeping the plates' relation with

Fig. 6.32 A view of the output and the atypical surface of the mined rock block obtained for the most favourable set of direction and revolutions number of the head body and plates with disk tools [1, 4]

Fig. 6.33 A view of: **a** flat plate solution with the wear signs on the plate side surface and disk tools, **b** new plate solution in the shape of flower petals [4]

disc tools rotations number to the number of the head body rotations within the range of 3–3.3. It was stated that the most favourable direction of mining for this configuration of rotations was the horizontal one from right to left from the depth of 0 to maximum. The maximal depth value should not exceed 15–20 cm for the suggested head solution. It was also possible to mine vertically from top to bottom and horizontally from left to right at the lowered speed of the head body movement. However, sample mining was more difficult.

The new solution of the plates was developed to increase the life period and decrease the wear of the plates and disk tools. Instead of a round plate, a plate in the shape of flower petals was suggested. The disk tools were mounted on the new plates so that their axis of rotation was tilted inwards by an angle of −5°. The new plate with disk tools is shown in Fig. 6.33b. To check the influence of mounting and the deflection angle of the rotational axis of the disk tool, new exchangeable holders of these tools and plates were made, allowing disk tools assembly at an angle of −5°, 45° and 90° relative to the surface of the plate. The models of these holders and a view of the plates with disk tools with these holders during field tests are shown in Fig. 6.34 [1].

The trials to excavate the concrete block with disk tools mounted at 45° and 90° to the surface of the plate surface ended with quick and catastrophic wear of the tool. The view of the destroyed tool mounted at an angle of 90°, compared to the tool fixed at an angle of −5° after about an hour of work, is shown in Fig. 6.35. The attempts to replace a disk tool with a smooth wedge by another, reinforced with sintered carbide inserts, have also been unsuccessful. The inserts were broken after a few minutes of the work. The view of the new solution of disk tools reinforced with sintered carbide inserts and view of fractures and breakage of carbide inserts is shown in Fig. 6.36.

During the last phase of the field tests, the mining head was mounted on a larger roadheader, weighing over 65 tones and equipped with a larger power electric motor—250 kW. Two additional hydraulic motors HS 0.8 to drive the mining head body were also included. This allowed for better, more stable operation of the mining head and the possibility of mining at greater depths [2]. The view of this roadheader and mining head before and during the trials is shown in Fig. 6.37.

During the tests, using disk tools made of different materials and by various methods, the efficiency of mining was checked. The disk tools had the same diameter—170 mm and edge angle of 45°. They were made of two types of tool steel, low-alloyed (GS42CrMo4) cast steel and ADI spheroidal cast iron. The disk tools manufactured from tool steel type NZ3 were full-hardened to a hardness of over 55 HRC. The tools from tool steel type 36HNM were surface-hardened and tempered. Tools made of cast steel and cast iron were machined after casting. After this mechanical treatment, it was noticed that especially tools made of cast steel had numerous castings defects—blisters and pores. These tools have been disqualified. These are shown in Fig. 6.38 [1].

The best results were obtained during concrete block mining with a more significant cutting depth for disk tools made from tool steel type 36HNM. After one hour of operation, no significant traces of wear were observed. In the case of disk tolls

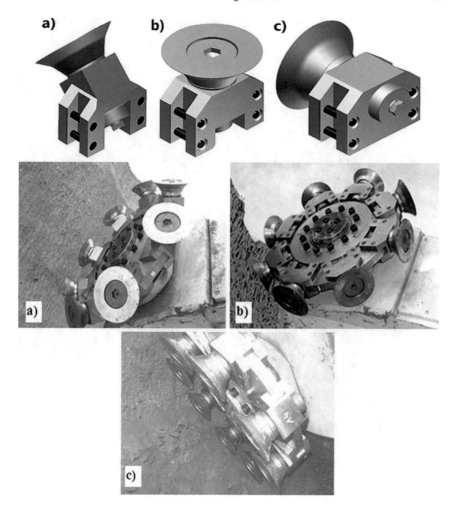

Fig. 6.34 Spatial models of new exchangeable holders of disk tools and a view of the plates with disk tools allowing mounting them at an angle of: **a** 45°, **b** 90°, **c** −5° relative to the surface of the plate according to [1]

made from tool steel NZ3, numerous edge breaks and cracking and breakage of the part of tools were noticed [2].

However, in the case of disk tools made of ADI cast iron, no breakage of the tool edge but its systematic abrasive wear was observed. The outside diameter of cast iron tools after about 30 min of work was decreased to a value of about 135–140 mm.

This diameter value did not allow further work. The wear view of tools made from tool steel NZ3 and ADI type cast iron is shown in Fig. 6.39. The reason for quick tool wear was probably the large cutting depth and big movement speed of the mining head. The combination of these parameters caused a heavy load on the tools.

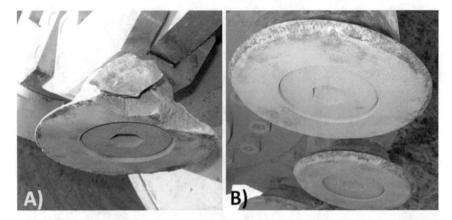

Fig. 6.35 A view of the disk tool edge wear: **a** disk tool mounted at an angle of 90°, **b** disk tool mounted at an angle of −5° [1]

Fig. 6.36 A view of disc tool reinforced with sintered carbide inserts and the traces of carbide inserts breakage

In the first part, a new conception of a mining head, in which the motion of tools will be forced and will cause mining of rock with tools along the complex trajectory, was described. This solution was developed in the Department of Mining, Dressing and Transport Machines, AGH University of Science and Technology, Cracow, based on an analysis of the world technique condition and results of own tests of rocks mining with asymmetrical disk tools of diameter D up to 160 mm. The mining method mode of this mining head solution allows the crossing of mining lines of individual disk tools and facilitates rock mining by breaking off rock furrows.

The next part describes the laboratory trials to choose the most convenient geometric parameters of the symmetrical mini-disk tools and the way of the mining process. The tests were conducted on the unique laboratory stand in two phases—first only for a single plate with mini-disk tools with a straight-motion mining trajectory— in the horizontal or vertical plane. The concrete and natural rock samples were mined.

Fig. 6.37 A view of new larger roadheader with new mining head with disk tools before and during the field tests according to [2]

Fig. 6.38 A view of defective disk tools made of cast steel with: **a** pores, **b** blisters [2]

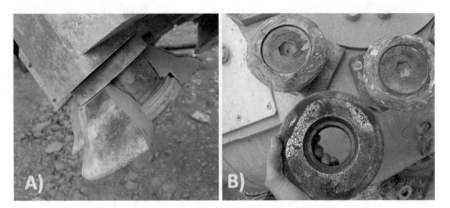

Fig. 6.39 A view of disk tool wear after field trials: **a** made from steel type NZ3, **b** made from ADI cast iron [2]

The most favourable working conditions for obtaining the lowest load were determined. The geometric parameters of the symmetrical mini-disk were also selected while pressing the single mini-disk tools into the rock sample.

The obtained results were checked on the laboratory stand for single mining tools testing during mining the concrete ring samples with a single plate with mini-disk tools with a complex-motion trajectory.

In the next part, based on the selected working parameters of the new mining head solution, the mining head 3D model was designed, and a prototype of the head was manufactured. The new mining head was prepared for assembly on the arm of medium type roadheader.

The field trials were performed on the large size concrete block to choose the most convenient working parameters of the prototype mining head—low load, extensive output graining, and small vibrations. Several parameters were checked, direction and number of rotations of the plates with disk tools and the mining head body, type of mini-disk tools, the direction of mining.

The suggested solution of mining head with asymmetrical mini-disk tools of complex motion trajectory proved its usefulness while compact rocks mining and can be used as an alternative for existing mining heads of roadheaders. A selection of proper configuration of direction and number of rotations of the head body and plates with disk tools is necessary to obtain the most favourable parameters of the mining operation—the large size of output graining, the low load of drive engines and low vibration levels.

The rotations of plates with disk tools should be in a clockwise direction, and the head body rotations in a counter-clockwise direction. The relation of the plates with disc tools rotations number to the head body rotations should be within the range of 3–3.3. It was stated that the most favourable direction of mining for this configuration of rotations was the horizontal one from right to left from the depth of 0 to maximum. The maximal depth value should not exceed 15–20 cm for the suggested head solution.

The type of used mini-disk tools was significant—the used material and manufacturing technology. The most convenient results were obtained for the disk tools manufactured from tool steel type 36HNM—surface-hardened and tempered.

References

1. Kotwica, K., et al.: Collective Work—Selected Problems of Mining, Transport and Dressing of Highly Cohesive Rocks, Part 2, p. 137. AGH UST publisher (2017) (in Polish)
2. Kotwica, K., et al.: Collective Work—Selected Problems of Mining, Transport and Dressing of Highly Cohesive Rocks, Part 3, p. 128. AGH UST publisher (2018) (in Polish)
3. Kotwica, K., Gospodarczyk, P., Stopka, G., Kalukiewicz, A.: The designing process and stand tests of a new solution of a mining head with disc tools of complex motion trajectory for compact rocks mining. Q. Mech. Control 29(3), 119–129 (2010)
4. Kotwica, K., Gospodarczyk, P., Stopka, G.: A new generation mining head with disc tool of complex trajectory. Arch. Min. Sci. 58(4), 985–1006 (2013)
5. Kotwica, K., Klich, A.: Machines and Equipment for Roadways and Tunnels Mining, p. 313. Instytut Techniki Górniczej KOMAG, Gliwice (2011) (in Polish)
6. Kotwica K., Książkiewicz K., Gospodarczyk: P.: Mining Head for Compact Rock Mining. Patent of the Polish Patent Office. No. PL 209326, Warszawa (2011)
7. Kotwica, K., Prostański, D., Stopka, G.: The research and application of asymmetrical disk tools for hard rock mining. Energies 14(7), 1–21 (2021) Art. No. 1826

Chapter 7
Summary and Final Comments

Abstract In the last chapter, the author presents some suggestions that can improve the work presented in the book, new cutting tools and mining head solutions. The first part describes the concept of the lubricated crown pick. It is the connecting of presented in Chaps. 4, 5 new crown pick and lubricated holder. This solution is additionally equipped with water spraying during the cutting process. The next part presents a proposal for a four-plate mining head with disk tools of complex trajectory. The comparing computer simulations of the trajectory of disk tools during cutting with a three-plate head and a new proposal of a four-plate head were carried out. The obtained results are more favourable for the new proposal of a four-plate head. In the third part, the proposal of the concept of a single drive for mining head with disk tools of complex trajectory was presented. In the analysed drive system, the use of an asynchronous electric motor was assumed for two drive systems of the head body and the plates with disk tools. The comparative simulation tests of the new drive system and original system with independent drives of the head body and discs, were also carried out. The last part describes the proposal of new materials and manufacturing technology of asymmetrical mini-disk tools that allow their lifetime to increase, based on the innovative Cambridge Engineering SelectorEdu Pack program. Three different types of alloy steel and heat treatment processes for obtaining needed strength parameters were proposed, considering the costs and mechanical properties of analysed materials.

Based on the performed laboratory, field and industrial tests, it may be concluded that implementing a new solution of the cutting tool in the shape of a crown, as well as new, lubricated holders, allows to increase the lifetime of the cutting tools in comparison with traditional tools and holders. The encouraging results of tests enabled the development of new design solutions of mining heads with new crown picks and lubricated holders for roadheaders. These works were carried out in the Department of Mining, Dressing and Transport Machines, AGH UST Cracow.

Based on currently obtained results, attempts to use mini-disk tools (up to 170 mm in diameter) on the roadheaders mining heads offer promising results. The suggested solution of asymmetrical disk tools for mining heads with tools of complex motion trajectory proved its usefulness while compact rocks mining and can be used as an alternative for existing mining heads of roadheaders.

Below are some suggestions that can improve the results of the work presented in the book on cutting tools and mining head solutions.

In the event of satisfactory results of tests, roadheaders with new mining heads can be successfully used for drilling short-distance communication tunnels in compact and hard rocks. This may reduce tunnel excavation costs by decreasing the purchase, use and repair of drilling machines.

7.1 The Concept of New Lubricated Crown Tool

Correct operation of conical picks guarantees their high durability and affects the durability of the milling elements with low energy consumption of the process, low dust and sparking. However, obtaining such a result requires correctly selected kinematic and geometric parameters of the mining head with the entire machine (roadheader) and geometric and material parameters of the cutting tools with the holders. It is impossible or does not provide the desired results in many cases.

The Department of Mining, Dressing and Transport Machines, AGH UST Cracow, proposed and developed the solutions of the unique crown picks as an alternative to conical picks, allowing for multiple revolutions of the tool in the holder during cutting [1]. Promising effects were also possible to obtain with special, lubricated holders for conical picks. The new solution of the lubricated crown tool has been submitted to the Polish Patent Office and is covered by patent protection No. PL 210151 [4].

Based on the research results on crown picks and lubricated holders, a concept of a new type of crown tool together with a lubricated holder was developed. It has been designed in such a way as to make the most of the possibilities offered by its constructional shape for the additional use of internal sprinkling, lubrication of rotating elements, reduction of material consumption of the solution and simplification of assembly. The cross-sectional and isometric views of the new crown tool design are shown in Fig. 7.1.

The new crown tool and holder includes a pin holder 1, a working part in the shape of crown 2, to which eight carbide posts 3 are attached in the upper part. The crown is attached to the holder with a fixing screw 5. A water spraying nozzle 4 is built on the axis of this screw. A choking nozzle 6 is also mounted in the pin holder, which allows lubrication of the collaborated surfaces of the holder and crown tool. Due to the spraying on the tool, pure water was used as the liquid.

Due to the design of the crown-shaped working part of the tool and the possibility of lubricating the cooperating elements of the pick and the holder with the liquid, it will be possible to obtain numerous rotations in the cutting process. This will affect the low and symmetrical wear of the tool, which will ensure stable working conditions of the knife. In addition, a water spraying nozzle mounted in the axis of the pick will reduce the mining hazardous—dust and sparking [1].

Fig. 7.1 A new concept of a
lubricated crown
tool—cross-section and
spatial model: 1 pin holder, 2
crown, 3 carbide post insert,
4 spray nozzle, 5 fixing
screw, 6 choke nozzle
according to Ref. [1]

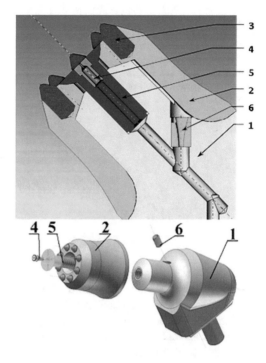

7.2   Proposal of the New Design of Four-Plate Mining Head with Disk Tools of Complex Trajectory

The development of assumptions and guidelines for a new head solution with disk tools with a complex motion trajectory is complex. They result from the unique concept of mining very compact rocks with disk tools and insufficient knowledge about the mining process in question. The consequence of the above is the need to examine some variables that determine the effectiveness of the mining process with an innovative mining head and its reliability.

Essential seems to perform further tests to elaborate the most favourable construction of the head and technology during drilling dog headings. Before further industrial trials, it is proposed to develop virtual models of disk tools and mining heads and to conduct computer simulation tests. The research results can allow achieving satisfactory results at much lower costs.

During field tests, it was proposed to use a follow-up control system for the head movement, which made it possible to correct its movement speed in the event of a temporary increase in the load on the drive system. The control system used for the field tests proved to be fully useful. However, this resulted in a reduction in cutting efficiency.

Therefore, it was proposed to modify the head structure to limit the load on individual disk tools without reducing the value of its speed of movement. It was

Fig. 7.2 Active surface of the disk—the contour of the cutting edges (marked in red) for the plate on which the tools are mounted at an angle of −5° to the axis of rotation of the plate [3]

also proposed to modify the design of the present head solution to eliminate two independent drives—separately for the body and plates with disk tools [3].

The obtained results of previous studies provided much information to develop guidelines and assumptions for further work on a new head solution with disk tools with a complex motion trajectory. Based on the observations made during the tests, it can be concluded that to ensure the proper operation of the disks (no abrasion of these disks by the mined rock mass), disk tools should be mounted so that the contour of their cutting edges (active surface) protrudes beyond the contour of the disk tool (Fig. 7.2). In the case of mounting the tools at an angle of −5° to the axis of rotation of the plate, it was found that the value of this contour (marked in red) must be at least 25 mm (suggested 30 mm or even more). This contour will ensure a sufficient cutting depth with a single tool without the disks rubbing against the surface of the excavated rock, even with disk tools with a certain degree of wear.

Adjusting the appropriate value of the speed of movement to the depth of cutting will reduce vibrations of the head during cutting and should ensure much more excellent durability of the device. Based on the analysis of the structure of the mining head used in the field tests and the observations made, it can be achieved by modernizing the structure of the head in question. Consider reducing the diameter of the head body or adding a plate with disk tools attached, which will increase the number of disk tools working per head revolution.

When developing the models of the modernized head solutions, it was assumed that the value of the drive power of the plates with disk tools would be a maximum of 150 kW, and the drive of the head body 100 kW. The complicated set of gears inside the currently used head solution (Fig. 7.3a) was replaced with a simpler one, consisting of conical gears (Fig. 7.3b). This solution reduced the diameter of the head body.

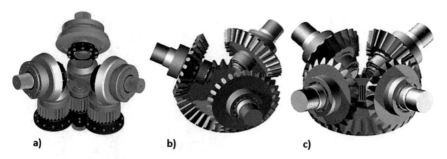

Fig. 7.3 A set of internal gears in mining head: **a** the current head design, **b** consisting of conical toothed wheels, **c** proposed conical for a four-disc head according to Ref. [3]

The advantage of such a solution is a simpler structure and higher permissible values of transmitted torques, the disadvantages are the multiplication of the rotational speed value and the necessity to ensure the coaxially of the plates with disk tools.

Subsequently, the possibility of replacing the three-plate head with disk tools with a four-plate version was considered. This solution has three main functional advantages [3]:

1. increasing the cutting efficiency, even up to 25%,
2. reduction of the amplitude of the values of loads acting on the head as a result of ensuring practically continuous support of the disks against the rock mass,
3. reduction of the cutting depth per one disk tool, resulting in its reduced load and reduced wear.

Considering that the additional fourth plate further complicates the gear system inside the head, it seems reasonable to replace the internal gear system of cylindrical gears with a bevel gear, whose main task is to incline the axis of the plates concerning the head axis. This type of transmission, shown in Fig. 7.3c, apart from smaller dimensions, enables the use of gear wheels with larger diameters and transmission of driving power with greater values.

In the case of a head with four plates, it was necessary to develop a simplified head body in a welded structure. The double-row planetary gear in the Paquera system with a gear ratio of 1:20 was proposed (Fig. 7.4), which allowed the desired value of the rotational speed and the value of the torque on plates with plates disk tools. The spatial model of a four-plate head with the Paquera planetary gear for the drive of plates is presented in Fig. 7.5 [3].

It is essential to choose the optimal value of the ratio of the rotational speed of the plates and the head body. The working trajectories of disk tools with the rotational speed of the plates equal to the speed of the head body are shown in Fig. 7.6a.

The trajectories of movement of disk tools are similar to circles. Such a solution should be suitable for not very compact rocks due to the large particle size distribution of the output graining. However, this is a disadvantageous variant of the work because

Fig. 7.4 Spatial model of a
two-row planetary gear in
the Paquera system [3]

Fig. 7.5 Spatial model of a
four-disk head with a
double-row planetary gear
moved to the arm of the head
boom [3]

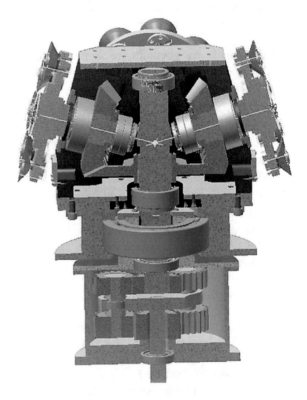

the plates drive is not used—only some tools effectively cut the rock, but their load
is impact-like, significantly shortening the tool life.

When the value of the rotational speed of the plates is 3 times higher than the speed
of the head body, the trajectory drawn by the disk tools ceases to be approximately
a straight line but takes the shape of a cycloid (Fig. 7.6b). This is an advantageous

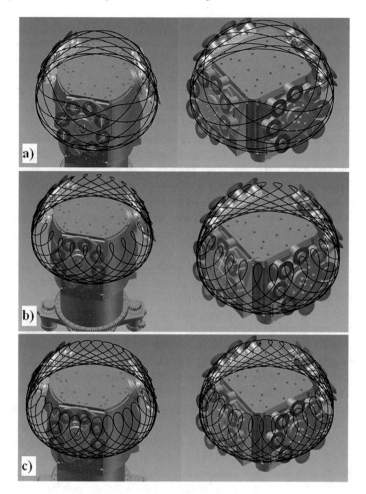

Fig. 7.6 Working trajectories of disk tools at the rotational speed of the plates: **a** equal to the speed of the head body, **b** equal to three times the speed of the head body, **c** equal to four times the speed of the head body according to Ref. [3]

arrangement; near the vertex of the cycloid, the rolling velocity vector of the disk tool is nearly parallel to the axis of rotation of the head. Such working conditions make the disk tool roll on the rock face, cutting off its successive layers. This results in a significant reduction of the loads acting on the head body. These mining parameters also allow the use of the entire circumference of each disk tool, the effective use of the plates drive and their more even load and wear.

It is also possible to determine the effective zone of their work for a given value of rotational speed, and thus—the effective depth of the head groove. This zone is approximately 140 mm for both the three-plate and the four-plate versions [3].

As the ratio of these velocities increases, the effective notch zone increases as the contact zone of the disk tool in contact with the rock increases in length more and more. As the value of the rotational speed of the plate's increases, the successive trajectories of the disk tools come closer together so that the amount of breakout material per disk is smaller and smaller, and thus the load on the head is also lower. In the case of compact and very compact rocks, the value of the rotational speed ratio, determined based on analyses of the kinematics of the cutting heads operation, is the most advantageous when the value of the rotational speed of the plates is four times higher than the rotational speed of the head body. The obtained working trajectories of disk tools are presented in Fig. 7.6c. The maximum effective cover for both of the presented head solutions is about 360 mm (about half the height of the mining head).

Further increasing the value of the rotational speed of the plates does not seem justified—the effective absorption increases to 500 mm, and at the speed ratio of 10:1, it will significantly reduce the granulation of the excavated material and more excellent abrasive wear of the tools.

During the operation of a cutting head equipped with three plates, there is a zone between the passage of two consecutive plates in which no disk tool is used. This arrangement is very disadvantageous because it causes considerable fluctuations in the amplitude of the loads on the cutting head. This work mode results in quick fatigue wear of the head elements and significant vibrations transmitted to the roadheader.

In the case of using four plates, the size of such a zone was significantly reduced. The design of this cutting head should show much lower load variability, which will increase the durability of both disk tools and other head components. The use of an additional fourth plate increases productivity, as it increases the value of the speed of movement of the head while maintaining the required spacing between the individual tools.

Figure 7.7 compares the trajectories of disk tools during cutting with a three-plate and four-plate head, at the same rotational speed of the head body and plates, and the speed of the mining head displacement. It is possible to notice clear densification of the lines of the detachable planes of the rock mass using an additional fourth plate (red motion path) [3].

7.3   Proposal of the Concept of Single Drive for Mining Head with Disk Tools of Complex Trajectory

Based on the results of laboratory and field tests with the mining head with disk tools of a complex motion trajectory, the number of guidelines for the prototype version of the mining head was accepted. The obtained results indicate the necessity to apply advanced model tests to verify the future, modified structure of the mining head. The results of simulation studies using the Multibody Simulation method, carried out in the field of dynamics of the head drive system, are presented below [3].

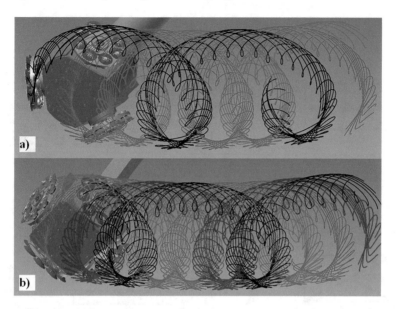

Fig. 7.7 Comparison of the trajectory of disk tools during cutting: **a** with a three-plate head; **b** with a four-plate head [3]

In the analysed drive system, the use of an asynchronous electric motor was assumed for two drive systems of the head body and the plates. It was assumed that the solution would be based on mechanical gears made of cylindrical gears.

The kinematic diagram of the drive system version, which allows the exact directions of rotation of the head hull and plates (clockwise or counterclockwise) to be obtained, is shown in Fig. 7.8a. It was assumed to reduce the diameter of the spacing of the disk tools, thus increasing the efficiency of the mining process. The diameter of the body may be reduced from 50 to 100 mm depending on other parameters of its construction. It should be of benefit to the rock mining process.

Necessary was made the modification of the last stage of the gearbox for obtaining rotation of the body and the plates with opposite directions to each other. Such a solution is shown in Fig. 7.8b. In this case, the gear wheel $z11$ should mesh with the bevel wheel $z12$ on the side opposite to the head face. A virtual model of the mining head was used for simulation tests, tested during field tests.

The simulation of the head operation was carried out by the task of kinematic inputs, i.e., a constant rotational speed at the input of the drive system. The simulation time was assumed to be 3 s. It was when one complete cycle of cutting a single plate took place. It was assumed that while the head was moving, the plates with disk tools would be loaded with a tangential force. The pressure force was not considered in the considerations. The tangential force assumed a value depending on the zone of its operation. In the middle cutting zone, it was 10 kN, while in the outer zones, it was 8 kN. The total value of the moment on one plate was 3000 Nm and was consistent with the average plate load value determined based on field tests. Simulation of the

a)

b)

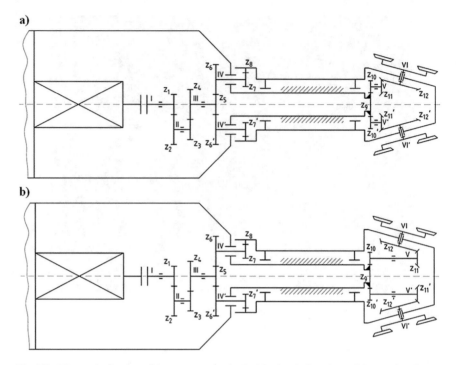

Fig. 7.8 Kinematic diagram of the concept of a single drive head, directions of the rotational speed of the head body and discs: **a** compatible; **b** opposite [3]

roadheaders mining system operation was carried out based on kinematic input with a constant rotational speed value at the level of the nominal engine speed of 1450 rpm. The simulation tests were performed for the same and opposite rotation directions of the plates and the head body [3].

As part of the simulation tests, comparative tests of the original drive system, i.e., a system with independent drives of the head body and discs, were also carried out. Based on the results of the simulation tests, a number of characteristics were obtained, the analysis of which allowed determining the sense of considering the modernization of the drive system concept for a new generation cutting head. The exemplary courses are shown in Figs. 7.9 and 7.10.

The graphs show the time characteristics of the power on the drive motors of the head body and the plates or the motor driving both elements. The occurrence of temporary peaks is related to the impulse appearance of forces on successive disk tools penetrating the rock mass.

For two independent drives, there are slight differences (in the order of several kilowatts) for the compatible and opposite directions of rotation of the plates and the head body. Consistent rotation directions mean that the course of the value of the required power is less dynamic. It has been noticed that the direction of their rotation with separated drives is irrelevant. The power distribution indicates the dominant

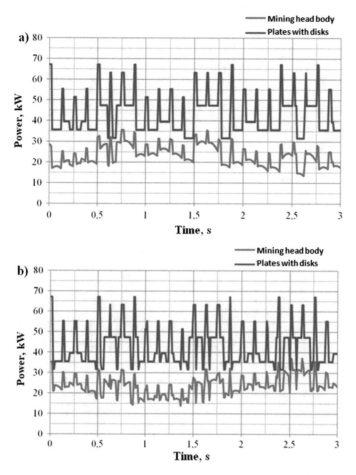

Fig. 7.9 Courses of the engine power demand for two drive sources, directions of the rotational speed of the head body and plates: **a** compatible; **b** the opposite according to Ref. [3]

(almost twice as large) share of plates with disk tools in the cutting process concerning the head body itself [3].

The analysis of the power demand in the case of a common drive source for the head body and plates (Fig. 7.10) showed that the configuration of rotations with a common drive is crucial. With opposite directions of rotation of the head body and plates, a significant increase in the demand for engine power occurs. The power value is comparable with the total power value for two independent drives. On the other hand, for compatible directions of rotation, the demand for engine power for one drive source is more than twice lower than in the case of separate drives [3].

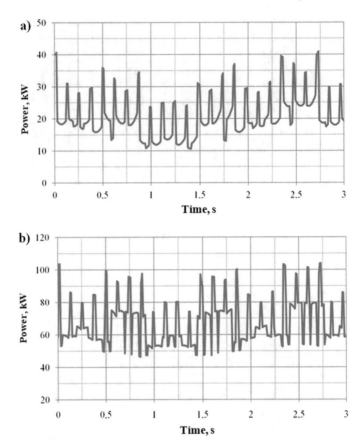

Fig. 7.10 Courses of the engine power demand for one drive source, directions of the rotational speed of the head body and plates: **a** compatible; **b** the opposite according to Ref. [3]

7.4   Proposal of New Materials and Manufacturing Technology of Asymmetrical Mini-Disk Tools

The mining of compact rocks with roadheaders with milling mining heads seems to be currently the most popular method. Today, attempts to use mini-disk tools (up to 170 mm in diameter) on the roadheaders mining heads offer promising results.

However, the biggest drawback in their case is durability. For disk tools lifetime increase, more attention should be paid to the way they are made and the material from which they are made and heat treatment.

For determining the materials that can be used for disk tools, it was proposed to use the innovative Cambridge Engineering SelectorEdu Pack program, available in an educational, free version for university employees and students. It uses an extensive database and can also be linked to its own database for specific applications [2].

An attempt was made to search for an alternative material solution for the disk tools used. Based on the analysis of mechanical methods of compact rock mining and the phenomena occurring in them, in particular the negative aspects of rock mining with asymmetrical disk tools, a formula was selected for the parameter of the functionality of a disk tool. It was assumed that the most important parameters in the case of disk tools are: fracture toughness $K^2{}_{1C}$, fatigue strength Z, yield strength R_e and hardness H. The suitability of the material for a disk tool should be selected after taking into account the product of these parameters, calculated according to the dependence (Eq. 7.1):

$$W_f(price) = K_{1C}^2 \cdot Z \cdot R_e \cdot H \qquad (7.1)$$

where:

$K^2{}_{1C}$—fracture toughness, (MPa· $m^{0.5}$),

Z—fatigue strength (10^7cycles), MPa,

R_e—yield point, MPa,

H—hardness, HV.

Based on the above dependence and the content of the CES EDU PACK database, the graph shown in Fig. 7.11 was created. In this graph, the abscissa axis represents the price of a given material, and the ordinate axis represents the value of the functionality parameter. Individual materials are marked with colours, such as alloy steel, tool steel, cast steel, cast iron, etc. The most suitable materials for disk tools are at the top of the chart. The selection tool was used to narrow down the material selection. The right side of the chart has been rejected on the assumption that the tool material should not be costly. Additionally, it can be noticed that the materials on the right and left sides of the graph do not differ significantly in terms of the values of the functionality parameter [2].

The selected section A of the diagram is enlarged as shown at the figure's top. From the selected materials, 3 proposals were selected for use in disk tools for hard rock mining. The following materials were proposed:

- Hy-Tuf low-alloy steel, hardening and tempering,
- Medium-alloy steel Fe-5Cr-Mo-V, hardening and tempering,
- AF1410 high-alloy steel.

The individual values of their most critical mechanical parameters have been collected and presented in Table 7.1. The type of treatment and thermal improvement was also proposed. The following heat treatment processes were recommended to obtain the values of the mechanical parameters presented in Table 7.1 [2]:

- AF1410 high-alloy steel.
- Hy-Tuf low alloy steel: normalizing annealing at 976 °C, austenitizing at 871 °C, heating to 204–316 °C and then hardening in air,

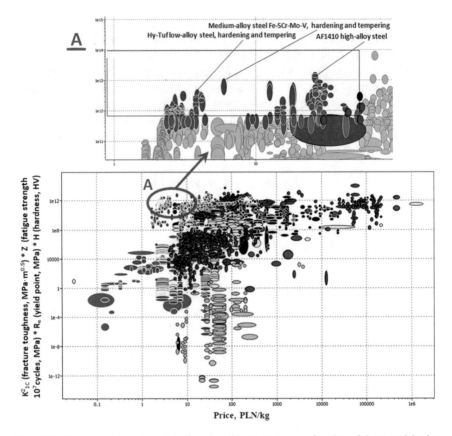

Fig. 7.11 Diagram of the value of the functionality parameter as a function of the material price prepared according to the program and the CES EDU PACK database [2]

- Fe-5Cr-Mo-V medium-alloy steel: material heated at 1000 °C, then air-hardened, high tempering at 580 °C for one hour,
- AF1410 high alloy steel: material heated at 857 °C, then oil-hardened, high tempering at 510 °C for 5 h.

7.5 Summary

Mining compact rocks with mechanical methods using arm roadheaders with heads with tangential rotary picks are currently the most popular method of drilling headings. However, it is connected with increased wear of cutting tools. It is possible to apply the high-pressure water jets assistance of the mining process with cutting tools

Table 7.1 Chosen values of mechanical parameters of selected steels

Parameter	Steel					
	Low-alloy		Medium-alloy		High-alloy	
	Min	Max	Min	Max	Min	Max
Price, [PLN/kg]	3.65	4.01	5.66	6.23	25.0	27.5
Density, [kg/m^3]	7790	7870	7740	7820	7790	7870
Young's modulus E, [MPa]	200	210	207	218	203	213
Yield point R$_e$ [MPa]	1280	1410	1380	1660	1480	1640
Compressive strength σ_c, [MPa]	1330	1470	1520	1790	1540	1700
Tensile strength σ_t, [MPa]	1520	1680	1660	1930	1620	1790
Hardness, [HV]	167	500	207	690	500	600
Fracture toughness K$^2_{1C}$, [MPa·m^{0,5}]	90	110	104	136	150	160
Fatigue strength for 10^7 cycles, [MPa]	683	835	662	869	650	750

to increase durability, safety and mining heads work efficiency. Due to fighting explosion and dustiness threats, the system with rear assistance is preferred. However, as demonstrated by performed and described tests, better results can be obtained by applying water assistance (lubrication) of tool holders, even with clean water. The method enables the increase of rotary tools rotation in the holder and allows obtaining lower and more even wear of tools edges.

Based on the performed laboratory, field and industrial tests, it may be concluded that implementing a new solution of the crown tool and lubricated holders allows an increasing lifetime of the cutting tools compared to traditional tools and holders. The encouraging results of tests enabled the development of new design solutions of mining heads with new crown picks and lubricated holders for roadheaders. The field trials have given satisfying results. In the case of industrial tests, the developed solutions require the development of a follow-up head control system to reduce the load on the mining tools and a lubricant fluid distribution system to the holders. These works were carried out in the Department of Mining, Dressing and Transport Machines, AGH UST Cracow.

The results of laboratory and field tests presented in the book constitute unique information regarding the impact of geometric parameters of disk tools and mining process parameters on the reaction force generated during mining. From a practical point of view, the presented test results can be input data for design purposes and can be used to select the necessary parameters of the mining heads. The relations between the disk tools individual geometric parameters, the mining process and the reaction forces, identified as a result of the conducted research, can be used to validate analytical and simulation mining models.

Preliminary tests have shown great potential for increasing mining efficiency by using the mining head with mini-disk tools. However, this requires selecting kinematic parameters that will not generate too large reaction forces on the mining head and the machine body. It will enable better control and steering of the mining process.

The conducted research and obtained results significantly broaden the scope of knowledge in mining with asymmetrical disk tools. The practical (utilitarian) aspect of described laboratory research work involves test results for selecting kinematic and dynamic parameters of a new type of mining head for roadheaders, equipped with disk tools. The research results on the prototype mining head confirmed the validity of the methodology used for testing with a single disk and plate with disk tools.

References

1. Kotwica, K., et al.: Collective work—selected problems of mining, transport and dressing of highly cohesive rocks, Part 1, p. 147. AGH UST Publisher (2016). (in Polish)
2. Kotwica, K., et al.: Collective work—selected problems of mining, transport and dressing of highly cohesive rocks, Part 3, p. 128. AGH UST Publisher (2018). (in Polish)
3. Kotwica, K., et al.: Collective work—selected problems of mining, transport and dressing of highly cohesive rocks, Part 4, p. 181. AGH UST Publisher (2020). (in Polish)
4. Kotwica, K., Kalukiewicz, A., Minorczyk, G., Gospodarczyk, P.: Mining tool especially for hard rock mining. Patent of the Polish Patent Office. No. PL 210151, Warszawa (2011)

Printed in the United States
by Baker & Taylor Publisher Services